THE SEARCH FOR
ORGANIC REACTION PATHWAYS

The search for organic reaction pathways

Peter Sykes, M.Sc., Ph.D., F.R.I.C
Fellow of Christ's College, Cambridge

A Halsted Press Book

John Wiley & Sons
New York

LONGMAN GROUP LIMITED
London
*Associated companies, branches and representatives
throughout the world*

© Longman Group Limited 1972

First published 1972

Published in the U.S.A.
by Halsted Press Division,
John Wiley & Sons, Inc.
New York

ISBN 0 470 84130 3

Library of Congress Catalog Card No 72 4192

Printed in Great Britain by
J. W. Arrowsmith Ltd., Bristol

954180

To The Chemistry Faculty
    The College of William and Mary
    With affection and respect

Contents

Foreword

It has been said that nothing in life is permanent except change. Never was this more evident than in Chemistry today. A student is faced at present with a rapidly increasing body of facts and with frequent revisions of the theories which are advanced to correlate and explain these facts. He asks, rightly, how many of these he should try to remember and how relevant are they for an understanding of the subject. Unless the facts and theories are taught in a way which excites his imagination and his desire to learn more about the subject, his interest in chemistry will wane rapidly.

This book by Dr Peter Sykes sets out to stimulate and maintain his interest in organic reaction mechanisms by showing how experimental data can be used to find out how organic reactions take place. As an experienced original research worker and a first-class teacher he starts with facts and with the simple principles needed to interpret these, avoiding the excessive amount of mathematical theory so loved by those authors who wish to impress their academic colleagues but who more frequently depress the students by whom the book will be used. The student is then guided through to an understanding of the possible mechanisms which can be inferred from the data, with frequent warnings against trying to infer too much.

This book is a fine example of a good teacher in action and by a good teacher I mean a person who knows his subject well, understands it and recognises what to teach and, even more important, what not to teach. The approach adopted encourages the student to want to learn, for when one assesses a teacher one should ask not so much what did he teach me, but did he teach me how to learn.

University College, London   R. S. NYHOLM

Preface

The initial motivation for this book came from a lecture to the annual meeting of the Association for Science Education in London in 1967. That lecture was called *Finding Out About Organic Reactions*, and it sought to describe the various sources from which the basic experimental data on reaction pathways could be obtained and, once obtained, how it could be fitted together so as to become meaningful in terms of connected molecular events. A number of people at the meeting were kind enough to say that they found such a synthesis stimulating and interesting, but that they had difficulty in finding the relevant information conveniently collected together. It seemed, therefore, that a relatively simple book on the subject might possibly be useful: this is an attempt at such a book.

In a number of ways this work is a complement to my *Guidebook to Mechanism in Organic Chemistry* (3rd Ed., 1970); the latter is essentially a didactic book in which an attempt is made to present the basic facts and situations of mechanistic organic chemistry—the very vocabulary of the subject—in as simple a context as possible. With such an objective, there was usually neither time nor space to dwell at any length on the translation of raw experimental data into relatively polished mechanistic ideas. The present work seeks to deal, therefore, with that very aspect of the subject.

So far as the pattern of the present book is concerned, the titles of the individual chapters are virtually self-explanatory and their contents do, I hope, provide a useful and functional way of grouping the material. To start with kinetic data and its interpretation seems not merely logical, but indeed inescapable, for this is the real origin of all our ideas about reaction pathways. The subsequent chapters are then arranged so that the relative sophistication of the arguments employed, and of the situations considered, could be gradually increased as they are traversed. In the last chapter a few further, and contrasted, examples are provided in which the techniques and arguments that have been employed previously are either carried a little further, or several of them are used together, to provide rather more detailed information about the pathway by which a particular reaction is believed to proceed. Having said that, it is perhaps necessary to emphasize that this is essentially a simple book: little or nothing beyond the most elementary ideas about organic chemistry are required in order to be able to read it, though careful reading of it may be required at least in the later parts.

As with the *Guidebook*, I am indebted to many people for ideas and suggestions made over the years; though the origins of not a few of them are now, I am afraid, forgotten. In more concrete terms, I am indebted to Miss Janet Thompson and Miss Christine Hacker for turning a labyrinthine manuscript into an orderly typescript; to Professor Melvyn D. Schiavelli for reading the manuscript in draft, and for making, in many diverting conversations, numerous useful suggestions; to Professor Sir Ronald Nyholm, F.R.S., a great chemical educator, for so kindly writing the Foreword; to my daughter, Helen, for reading the proofs to meet a deadline; to my wife, Joyce, for reading the manuscript, final typescript and proofs, for making a multitude of helpful suggestions, for insisting that I wrote what I actually meant and not something just vaguely like it, and for being a tower of strength, support, help and comfort at all times. Finally, I am especially indebted to the College of William and Mary, through the agency of my old friend Professor Sheppard Y. Tyree Jr., Chairman of the Chemistry Department, for the invitation to spend the academic year 1970–1971 there as a visiting research professor. This gave me the freedom from my usual commitments that made the writing of this book possible; it also proved to be one of the happiest years of my life.

Williamsburg, Virginia   PETER SYKES
Cambridge, England
August 1970–September 1971

Note added in proof:

In the final stages of proof correction the news came of the tragic death of Sir Ronald Nyholm: a great chemist and a very great human being. He was always the same: effective, helpful and never other than kind; it was a positive tonic to be in his company. Sadly, we shall not see his like again.

1
Kinetic data and its interpretation

STOICHIOMETRY AND RATE OF REACTION

An erroneous belief—singularly difficult to convince oneself out of—
is that the stoichiometry of a reaction, however lovingly determined,
can tell us anything at all about the rate at which starting materials are,
under given conditions, converted into products, or about the actual
pathway by which this conversion takes place. The erroneous nature of
the belief is particularly difficult to accept in a simple case such as

$$RBr + {}^{\ominus}OH \longrightarrow ROH + Br^{\ominus}$$

where one instinctively assumes that the reaction will proceed in one
step *via* a two-body collision between molecules of the reactants (i.e. an
elementary reaction), and that the rate of reaction will thus be simply
related to the concentrations of RBr and ${}^{\ominus}OH$. This may indeed be
what actually happens (e.g. $MeI + {}^{\ominus}OEt$, *cf.* p. 17) but, equally, it may
not (e.g. $Me_3CCl + H_2O$:, *cf.* p. 14), and one is never justified in

1

assuming such a one-step, two-body collision purely on the basis of the reaction's stoichiometry. This is shown particularly clearly in the acid-catalysed iodination of acetone (1) (*cf.* p. 28)

$$CH_3COCH_3 + I_2 \xrightarrow{\;H^\oplus\;} CH_3COCH_2I + H^\oplus I^\ominus$$
(1)

where the rate of reaction is found to be independent of the concentration of iodine, despite the fact that the latter appears in the equation and is incorporated into the reaction product: the iodination therefore clearly cannot proceed by a simple one-step pathway involving acetone and iodine.

This lack of any necessary relation, between a reaction's stoichiometry and the actual molecular encounters through which it proceeds, comes to seem a little less unreasonable in the light of an example such as:

$$\underline{2}MnO_4^\ominus + \underline{10}Fe^{2\oplus} + \underline{16}H^\oplus \longrightarrow 2Mn^{2\oplus} + 10Fe^{3\oplus} + 8H_2O$$

The very idea of a $2 + 10 + 16 = 28$-body collision—as required by a one-step pathway— is statistically absurd, and the overall reaction must, inescapably, proceed *via* a series of successive steps, i.e. a *complex reaction*. Each of these successive steps is likely on statistical grounds to involve only a very small number of molecular (or ionic, or atomic) species—usually not more than two—though some of these reacting species may, in later steps, be composites, e.g. protonated forms etc., elaborated in earlier stages. The rate of the overall process will thus be defined by the relative rates of these individual steps, and it is common to find that one of them is markedly slower than the others; such a slow step constitutes a 'kinetic bottleneck', and thereby sets a limit to the rate of the overall reaction: it is therefore often referred to as the *rate-limiting step* (*cf.* p. 23). It thus becomes increasingly apparent that in order to establish the relationship between the concentration of any one, or all, of the starting materials (or products) and the rate of reaction, there is no alternative to actual experiment.

HOW FAR AND HOW FAST?

Further, there is found to be no necessary connection between the completeness of a reaction and its rate. Because a reaction is found to proceed so that starting materials are converted essentially 100 per cent into products, this gives no guarantee whatever that the conversion will proceed at an acceptable rate under readily accessible conditions: clearly, different factors must be involved in determining *how far* a reaction will go to completion, and *how fast* it will do it. This is seen very readily in, for example, the air oxidation of cellulose (2),

$$(C_6H_{10}O_5)_n + 6nO_2 \longrightarrow 6nCO_2 + 5nH_2O$$
(2)

where though the reaction can be shown to proceed essentially to completion, it is nevertheless possible to read a newspaper (very largely composed of cellulose) in air—or even in an oxygen tent!—for very long periods of time, without it bursting into flames or gradually fading away to gaseous products ($CO_2 + H_2O$). The rate of reaction (how fast?) is very low indeed at room temperature despite an equilibrium (how far?) that is attempting to proceed virtually to completion over on the right-hand side.

(a) The standard free energy change, $\Delta G°$

It has been found that the factor which determines how far a reaction will go in converting starting materials into products is the difference in stability between products and starting materials: the more stable the products are compared with the starting materials, the further the reaction is found to go towards completion. Stability, in this context, is however, found to involve more than merely the simple energy difference between products and starting materials as measured by the heat (more exactly, *enthalpy*) of reaction, ΔH (written as $-x$ kiloJoules, or kilocalories, per mole for an *exothermic* reaction in which heat is evolved, i.e. *lost* by starting materials on conversion into products). This is reflected in the fact that highly exothermic reactions (large negative values of ΔH), in which the products have a considerably lower heat content ($\approx$ energy level) than the starting materials, are not necessarily found to go further towards completion than those reactions for which ΔH has a smaller negative value, nor even than those for which ΔH has a positive value (*endothermic* reactions): no simple correlation is found between K, the equilibrium constant, and $-\Delta H$ (i.e. the heat evolved), so some additional factor must clearly be involved.

That this should be so is a corollary of the Second Law of Thermodynamics which is concerned essentially with probabilities, and with the inexorable tendency of ordered systems to become disordered: a measure of the degree of disorder of a system being provided by its *entropy*, S. In achieving their overall most stable state, systems are found to tend towards *minimum* energy (actually enthalpy, H) and *maximum* entropy (disorder or randomness, S). A measure of their relative stability must thus embrace a satisfactory compromise between H and S, and is provided by the *Gibb's free energy*, G, which is defined by

$$G = H - TS$$

where T is the absolute temperature. The free energy change during a reaction, at a particular temperature, is thus given by

$$\Delta G = \Delta H - T\Delta S$$

and it is found that the free energy change on going from starting materials to products under standard conditions (i.e. for 1 mole and at 1 atmosphere pressure), ΔG°, is related to the equilibrium constant, K, for the reaction by the relation:

$$-\Delta G^\circ = 2 \cdot 303 \, RT \log_{10} K$$

The larger the *decrease* in free energy on going from starting materials to products, i.e. the larger the *negative* value of ΔG°, the larger the value of K and the further over to the right the equilibrium will therefore lie, in favour of products. Thus for a reaction in which there is no free energy change ($\Delta G^\circ = 0$), $K = 1$ corresponding to 50 per cent conversion of starting materials into products. Increasing *positive* values of ΔG° imply *rapidly decreasing* fractional values of K (the relationship is a *logarithmic* one) corresponding to extremely little conversion into products, while increasing *negative* values of ΔG° imply correspondingly rapidly *increasing* values of K; thus a ΔG° of $-42\text{kJ}/$ mole corresponds to an equilibrium constant of $\approx 10^7$, and essentially complete conversion of starting materials into products.

(b) The free energy of activation, $\Delta G^{\ddagger}$

But what of the factor that determines how fast a reaction will go, given—as is demonstrably the case—that it is *not* ΔG°? The pathway, in free energy terms, for an elementary reaction is seldom if ever a mere run downhill, even though it may have a large negative value for ΔG° (Fig. 1.1):

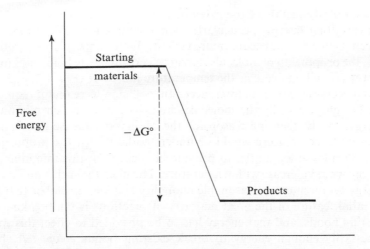

Fig. 1.1

If the above did represent a reaction's energetic pathway we should expect essentially every collision between reactant molecules to lead to product formation, whereas in fact the number of successful (i.e. leading to product formation) collisions can be shown, for the great majority of reactions, to be only a very small proportion indeed of all those taking place. Something other than mere encounter is clearly necessary if the collision is not to be simply an elastic one. The need for this extra requirement we might perhaps have forecast from the well known observation that the rates of chemical reactions, irrespective of whether they are exothermic or endothermic, increase with temperature.

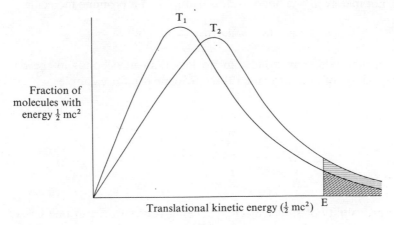

Fig. 1.2

The simplest effect that a rise in temperature can have on molecules is to increase their energy, particularly their kinetic energy; collisions between them will thus become more violent. By Boltzmann's distribution law, the proportion of molecules having a certain, fairly high, minimum energy (E) will increase as the temperature increases (Fig. 1.2, p. 5).

If a necessary criterion for a successful collision were that it required the bringing to it, by the molecules involved, of a certain minimum energy, say E, then the reason for the observed increase in reaction rates with temperature would be understandable; for the proportion of molecules able, jointly, to provide not less than this minimum of energy would increase with temperature. The need for such a minimum energy requirement is reasonable enough when we remember that an essential feature of the great majority of reactions is the breaking of existing bonds, and that energy has to be provided to effect this: not necessarily enough energy to break existing bonds *completely*, but enough at least to increase their vibrational energy, and so to begin stretching them preparatory to breaking.

However, just as ΔH° alone did not correlate with K, the equilibrium constant, a simple energy term is not of itself sufficient wholly to explain the barrier that has to be overcome in converting starting materials into products: here, too, a probability term is also involved. Not only that, this probability factor can be involved in two essentially different ways: in an energy context and in a spatial context. The energy context arises because the great majority of organic molecules are polyatomic and thus contain a number of, often many, different bonds. When a non-elastic collision takes place between such reacting species, the kinetic energy can thus be taken up, as vibration leading to stretching, not merely in one bond as, for example, in a bromine molecule

$$Br\!-\!Br \xrightarrow{\;E\;} Br\cdots Br \longrightarrow Br\cdot + \cdot Br$$

but in many different bonds: in no less than nine bonds in the decomposition of cyclopropane (3) to propene (4), for example:

The probability of sufficient energy being concentrated in one C—C bond, to the partial exclusion of the others, so as to effect its stretching, preparatory to fission, is thus relatively unlikely, and this imposes a further limitation—in fact a probability or entropy limitation—on the collision being successful, even though sufficient total energy may be brought to it.

A further probability factor, this time in a spatial context, is also involved as the majority of organic molecules are relatively complex in structure, as compared with simple diatomic molecules such as Br—Br, H—Cl etc. This means that a collision between reacting molecules needs to be not only sufficiently energetic but also satisfactorily oriented, if it is to have a successful outcome in terms of leading to the expected products. Thus in the base-induced hydrolysis of a primary alkyl bromide with a straight chain of, for example, five carbon atoms (1-bromopentane, 5), its collision with a hydroxyl ion has to be oriented such that contact takes place with the terminal carbon atom carrying bromine—if hydrolysis is to take place—collision with the other four carbons atoms just will not do:

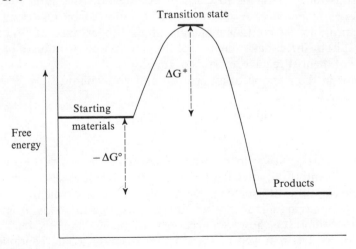

This obviously imposes a specific limitation on any given collision having a successful outcome.

The barrier that has to be overcome in converting starting materials into products can be represented, for an elementary reaction, by an energy profile such as Fig. 1.3:

Fig. 1.3

$\Delta G^{\pm}$ is usually described as the *free energy of activation*, and is defined by

$$\Delta G^{\pm} = \Delta H^{\pm} - T\Delta S^{\pm}$$

where $\Delta H^{\pm}$ is the enthalpy of activation—the bond stretching term— and $\Delta S^{\pm}$ is the entropy of activation—the probability term. $\Delta G^{\pm}$ is associated with k, the rate constant for a reaction—the larger the value of $\Delta G^{\pm}$, the smaller the value of k, and the slower the reaction (see below, p. 172)—in a way similar to that in which $\Delta G°$ is associated with the equilibrium constant, K, for a reaction (p. 4). $\Delta H^{\pm}$ above is closely related to, though not actually synonymous with, the more familiar *activation energy*, E (*cf.* p. 20).

The point of maximum free energy in the conversion of starting materials into products is commonly referred to as the *transition state*; this is not a discrete, molecular species, but a unique arrangement of reacting entities in which existing bonds may be partly broken and new bonds partly formed. Even when the transition state has been attained product formation is not assured, for the transition state may still revert to starting materials as an alternative to being transformed into products.

EXPERIMENTAL KINETIC MEASUREMENTS

Having established the non-dependence of reaction rates on stoichiometry or on $\Delta G°$, and demonstrated the necessity for their experimental investigation, it remains to consider exactly how the latter may be achieved in practice. A reaction can be followed by observing the disappearance of starting materials or the appearance of products (or both) by direct determination of the actual concentration of one or more of them at regular intervals of time.

Thus in the reaction of an acid (6) and an alcohol (7) to form an ester and water

$$\underset{(6)}{MeCO_2H} + \underset{(7)}{EtOH} \longrightarrow MeCO_2Et + H_2O$$

perhaps the easiest way to follow the reaction is to determine the residual acid concentration, after the lapse of successive intervals of time, by simple titration with standard base. This is most conveniently carried out by removing aliquots, after the various time intervals, from the main solution. It is, however, necessary to 'quench' the reaction in the aliquot as it is removed or the original reaction would continue to some extent during the subsequent titration, with consequent uncertainty about the exact time interval that has elapsed since the beginning of the overall reaction. Such quenching may be achieved in this case by running the aliquot directly into a large excess of water before titration takes place.

For every molecule of acid consumed in the reaction a molecule of alcohol will also be used up, and if acid and alcohol were initially present in equi-molar proportions then their concentrations will be found to change with time at the same rate and in the same way. We cannot, however, assume that twice as much acid (and alcohol) will be used up after, for example, 100 minutes as had been used up after 50 minutes; the rate of reaction is normally found to vary with time, depending—among other things—on the concentration of starting material (or materials) that has not yet undergone reaction. We have already seen (p. 5) that reaction rates are markedly influenced by temperature, so in making experimental kinetic measurements it is vital that the temperature is kept constant by allowing the reaction to take place in a thermostatically controlled constant temperature bath: the temperature is commonly maintained to within not more than $\pm 0 \cdot 1°$.

Rates of reaction can be established by means other than simple chemical determination of the concentration, after various time intervals, of residual starting materials (or of accumulating products). Thus in the conversion of the diazonium salt (8) into the phenol (9) in aqueous solution

$$ArN_2^{\oplus} + H_2O \longrightarrow ArOH + N_2 + H^{\oplus}$$
$$(8) \qquad\qquad\qquad (9)$$

the volume of gaseous nitrogen evolved may be measured directly with a burette or a gas syringe, i.e. at constant pressure. Indeed, any property of the reaction mixture that varies continuously during the course of a reaction, and whose relation to concentration of starting materials or products can be established (*cf.* p. 11), may be employed to determine its rate.

Such direct physical methods include measuring a change in total volume of the reaction mixture (with a dilatometer), as in the base-catalysed decomposition in aqueous solution of diacetone alcohol (10) to acetone (11):

$$Me_2C(OH)CH_2COMe \xrightarrow{\ominus OH/H_2O} 2Me_2CO \qquad \text{(volume increases)}$$
$$(10) \qquad\qquad\qquad (11)$$

Measuring a change in optical properties may also be employed; as, for example, the use of optical rotation—to follow the acid-catalysed hydrolysis of sucrose (12) to a mixture of glucose (13) and fructose (14):

$$(+)C_{12}H_{22}O_{11} + H_2O \xrightarrow{H^\oplus} (+)C_6H_{12}O_6 + (-)C_6H_{12}O_6$$
$$\text{(12)} \qquad\qquad\qquad \text{(13)} \qquad\quad \text{(14)}$$

Glucose (13) has only a small dextro ($+$) specific rotation value, while fructose (14) has a large laevo ($-$) value; as the reaction proceeds the initial ($+$) optical rotation of the solution (due to sucrose) is thus found to decrease, to become zero, and finally to become ($-$) (($-$) fructose > ($+$) glucose) as the reaction nears completion.

Other variables of the reaction mixture that have been employed include viscosity, refractive index, electrical conductivity, and spectroscopic properties (u.v., i.r., n.m.r. etc.). The latter in, for example the reaction of carbonyl compounds with amine derivatives:

i.r. (in D_2O):

$$\underset{\substack{| \\ CO_2^\ominus Na^\oplus \\ \nu_{max.}\ 1710\ cm^{-1}}}{MeC=O} + NH_2OH \longrightarrow \underset{\substack{| \\ CO_2^\ominus Na^\oplus \\ \nu_{max.}\ 1400\ cm^{-1}}}{Me-C=NOH} + H_2O$$

u.v. (in H_2O):

Here the characteristic absorption of $C=O$ in the starting material is normally found to disappear very rapidly indeed, to be replaced only slowly by the characteristic absorption of $C=N$ in the product, indicating the formation of an intermediate (*cf.* p. 82). Finally, if any of the reactants, or products, are coloured, the reaction may be followed by the waning or waxing, respectively, of colour in the solution (measured in a colorimeter); as, for example, the fading of colour (of iodine) in the acid-catalysed iodination of acetone (15) in aqueous solution (*cf.* p. 28):

$$CH_3COCH_3 + I_2 \xrightarrow{H^\oplus} CH_3COCH_2I + H^\oplus I^\ominus$$
$$\text{(15)}$$

One of the obvious advantages of using such physical methods is that they allow of direct measurements on the reaction solution without the necessity of removing aliquots after varying time intervals, with all the possibilities for error to which that may give rise. There is, however, the potential disadvantage that the property actually measured may not necessarily be linearly related to the concentration of one of the reactants or products. If such is the case, however, it is normally quite a simple matter to obtain a calibration curve by separately plotting values of the property being observed against the corresponding, known mixtures of reactants/products.

RATE/CONCENTRATION RELATIONSHIPS

Having devised experimental methods for determining the rate of a reaction it remains for us to establish, from our 'raw' kinetic data, the exact form of the relationship between the experimentally determined reaction rate and the concentration of one or more of the reactants (or products). This involves the conversion of titration figures, volumes of gas evolved, optical rotation data, electrical conductivity measurements, spectroscopic data, colour intensities etc. into concentration terms, either directly by calculation or *via* a separately determined calibration curve as described above.

For the essentially irreversible reaction

$$A + B \longrightarrow C + D$$

plotting [A] or [B], or plotting [C] or [D], against time yields typical curves such as Fig. 1.4 or Fig. 1.5, respectively,

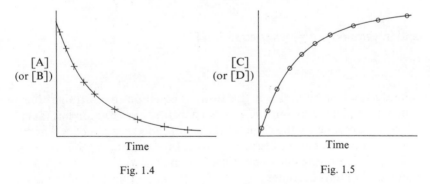

Fig. 1.4   Fig. 1.5

in which the rate of reaction at a particular concentration is given by the slope of the curve at that concentration. The necessary data can thus be obtained by the careful drawing of successive tangents to the curve at points corresponding to a number of different concentrations; but this does, in practice, impose considerable manipulative difficulties. These may usually be avoided by *assuming* a particular relationship between the concentration of one—or more—of the reactants (or products) and the rate of reaction, as measured by the rate of disappearance of that reactant or reactants or the rate of appearance of that product or products, i.e. the change in its (their) concentration with time, t. The experimental data may then be fitted into the particular mathematical equation required by the assumed relationship, and if the fit is a very close one then the original assumption is deemed to be justified.

(a) First order reactions

If we *assume* for the generalized reaction above that its rate, at constant temperature, depends on the concentration of *one only* of the reactants, say A, then we can write:

$$\text{Rate} = \frac{-d[A]}{dt} \propto [A]$$

i.e. $\dfrac{-d[A]}{dt} = k[A]$

In the above the constant k is referred to as the *rate constant*, and the minus sign indicates that the concentration of A is decreasing as the reaction proceeds, i.e. as t increases. The above differential equation is not particularly easy to use in this form and is more convenient when integrated. Thus if the concentration of A at time t is expressed as c, we have

$$\frac{-dc}{dt} = kc$$

and integration of this gives equation [1]

$$-\ln c = kt + z \tag{1}$$

where z is a constant: this is the form of equation for a straight line. z may be evaluated by putting $t = 0$, i.e. before any reaction has taken place, then c will have the value of the initial concentration of A that we arranged ourselves at the beginning of the experiment, and that we may call c_0. Substituting into equation [1] thus gives $z = -\ln c_0$, and equation [1] thus becomes

$$\ln c = -kt + \ln c_0 \tag{2}$$

or, converting to ordinary logs,

$$\log_{10} c = \frac{-kt}{2 \cdot 303} + \log_{10} c_0 \tag{3}$$

and this may, in turn, be rearranged to give:

$$k = \frac{2 \cdot 303}{t} \log_{10} \frac{c_0}{c}$$ [4]

In either differential or integrated forms such relations are known as *rate equations* or *rate laws*.

Then, provided our original assumption that reaction rate $\propto$ [A] only is valid, one or other of the varying forms of the integrated equation should be found to fit our experimental data closely. The equations may be tested either by substituting the known c_0 and corresponding c and t values in [4] and seeing if each pair of values gives approximately the same numerical value for k, or by plotting $\log_{10} c$ against t and seeing if a straight line is obtained (Fig. 1.6):

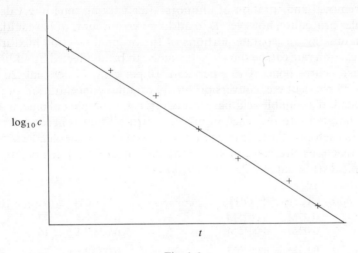

Fig. 1.6

If a straight line is obtained the intercept on the $\log_{10} c$ axis will be $\log_{10} c_0$ and the slope of the straight line will be $-k/2 \cdot 303$, from which an 'averaged' value of k, the rate constant, may thus be obtained; it will here have the dimensions of time^{-1}. A reaction for which equations such as [4] hold is said to be a *first order reaction*. The order of a reaction is defined formally as the sum of the powers of the concentration terms that occur in the differential form of the rate law; in this case there is only one concentration term anyway and it appears as [A], i.e. it is raised to the power one; hence the reaction is first order, but see below for other, more complex orders. If, however, the calculated values of k

increase or decrease progressively, or if the plot of $\log_{10} c$ against t is not a straight line, then our original assumption, that rate $\propto$ [A], is clearly not valid, and this cannot be a first order reaction. It should be emphasized that it is generally necessary to follow a reaction kinetically to not less than ≈ 60 per cent of completion if reliable estimates of its order are to be obtained; though there are exceptions to this.

A simple reaction that may readily be established experimentally as first order is the solvolysis in aqueous acetone of 2-chloro-2-methyl-propane (16):

$$Me_3C\text{—}Cl + H_2O \longrightarrow Me_3C\text{—}OH + H^{\oplus}Cl^{\ominus}$$
(16)

Acid is produced as the reaction proceeds and its formation can be followed and the reaction rate thereby determined, in the usual way, by the removal and titration of aliquots after varying time intervals. A simpler procedure, however, is to add known amounts of base initially when making up separate portions of the original reaction mixture— the reaction rate can be shown, separately, to be unaffected by addition of base—corresponding to 5 per cent, 10 per cent, 15 per cent, 20 per cent, 25 per cent etc. conversion of the original concentration of (16) into acid. If a suitable acid/base indicator, e.g. bromphenol blue, is also incorporated into the reaction mixture then all that is necessary is to note in each case the time from mixing before the colour change occurs (the reactions are, of course, conducted at constant temperature). A typical set of data (at 25°) is as follows:

Expt. no.	$[RCl]_0$	$[^{\ominus}OH]$	% conversion	$\therefore$ [RCl]	t(secs)	$10^3 k_1$
1a	0·03M	0·0015M	5	0·0285M	17	
1b	0·03M	0·0015M	5	0·0285M	16	2·96
2a	0·03M	0·0030M	10	0·0270M	34	
2b	0·03M	0·0030M	10	0·0270M	34	3·06
3a	0·03M	0·0045M	15	0·0255M	50	
3b	0·03M	0·0045M	15	0·0255M	55	3·15
4a	0·03M	0·0060M	20	0·0240M	78	
4b	0·03M	0·0060M	20	0·0240M	80	2·83

The first order rate constant, k_1, may readily be calculated by use of equation [4] and it will thus be seen that, at least over the range of 5–20 per cent conversion, this is substantially constant. The 'averaged' value for k_1 obtained by plotting $\log_{10} c$ against t (*cf.* Fig. 1.6) was found to be $2·99 \times 10^{-3}$ sec^{-1}.

(b) Second order reactions

Our original, essentially irreversible, reaction

$$A + B \longrightarrow C + D$$

may well turn out not to be first order; in which case it is very likely to be a *second order reaction*, which are indeed much the more common. In the general case of a second order reaction the rate of reaction will depend on the concentration of *both* A and B, and we can write, at constant temperature:

$$\text{Rate} = \frac{-d[A]}{dt} \propto [A][B]$$

i.e. $\quad \dfrac{-d[A]}{dt} = k[A][B]$

If, to simplify matters, we carry out the reaction with equimolar concentrations of A and B (i.e. $c_0^A = c_0^B$) and if the concentration of A at time t is expressed as c, then the concentration of B at time t will be c also and we will have:

$$\frac{-dc}{dt} = kc^2$$

Integration of this differential equation gives the equation [5]

$$\frac{1}{c} = kt + l \qquad\qquad\qquad [5]$$

in which l is a constant. l may be evaluated by putting $t = 0$, when l will thus be $1/c_0$ and [5] thus becomes:

$$\frac{1}{c} = kt + \frac{1}{c_0} \qquad\qquad\qquad [6]$$

$$or \quad k = \frac{c_0 - c}{c \times c_0} \cdot \frac{1}{t} \qquad\qquad\qquad [7]$$

The validity of the second order assumption may be tested—as with the first order assumption above—either by substituting the known c_0 and corresponding c and t values in [7] and seeing if each pair of values gives approximately the same numerical value for the second order rate constant, k (with the dimensions: concentration^{-1} time^{-1}), or by plotting $1/c$ against t and seeing if a straight line results; if it does, its slope will be k and the intercept on the $1/c$ axis will be $1/c_0$.

A realized straight line relationship establishes that the reaction is, overall, second order but it does not necessarily establish that this arises from its being first order in A and first order in B, i.e. that the rate equation is:

$$\text{Rate} = k[\text{A}][\text{B}]$$

This is so because we started, for simplicity's sake, with equimolar concentrations of A and B (both c_0) leading to the same differential rate equation

$$\frac{-\mathrm{d}c}{\mathrm{d}t} = kc^2$$

as would be obtained from the rate laws:

$$\text{Rate} = k[\text{A}]^2 \qquad \text{i.e. second order in A}$$
$$or\ \text{Rate} = k[\text{B}]^2 \qquad \text{i.e. second order in B}$$

The question thus arises as to how we can tell which of the three possible rate laws is actually operating in a reaction that we have shown to be second order overall? This may be done by arranging that the initial concentrations of A and B are different, i.e. not equimolar, so that $c_0^A \neq c_0^B$. The concentrations of A and B at time t, c^A and c^B respectively, will now be different from each other also, but it is only necessary to determine one of them *experimentally*, e.g. c^A, as an equimolar amount of B will have been consumed during the same time interval, and c^B can thus be calculated for any value of t from the experimentally determined c^A and the known c_0^B. Thus if the decrease in concentration of A from zero time (start of reaction) to time t is $x\ (= c_0^A - c^A)$, then $c^B = c_0^B - x$. The differential form of the second order rate expression now becomes

$$\frac{-\mathrm{d}c^A}{\mathrm{d}t} = kc^A c^B$$

and integration of the equation, and conversion from ln to $\log_{10}$ gives [8]:

$$\log_{10}\frac{c^B}{c^A} = \frac{(c_0^B - c_0^A)}{2 \cdot 303}kt + \log_{10}\frac{c_0^B}{c_0^A} \qquad [8]$$

Plotting $\log_{10} c^B/c^A$ against t will then give a straight line if the second order reaction is indeed first order in A and first order in B, and the second order rate constant, k, may be evaluated from measurement of the slope of this line which will be equal to $(c_0^B - c_0^A)k/2 \cdot 303$.

A simple reaction that may readily be established experimentally as first order in each of the two reactants, i.e. second order overall, is the reaction of iodomethane (17) with sodium ethoxide (ethoxide ion, 18) in ethanolic solution:

$$EtO^\ominus + MeI \longrightarrow EtOMe + I^\ominus$$
$$(18) \quad (17)$$

The reaction may be followed, at constant temperature, by removal of aliquots after various time intervals, quenching the reaction by dilution with water, and determination of the residual concentration of $EtO^\ominus$ by titration with standard acid ($\approx 0\cdot050M$ $HClO_4$); the initial concentration of $EtO^\ominus(c_0{}^A)$ was $0\cdot0404M$ and of $MeI(c_0{}^B)$, $0\cdot0576M$. A typical set of data (at 21°) is as follows:

t(min)	$HClO_4$(ml)	c^A	c^B	c^B/c^A	$\log_{10} c^B/c^A$
0·00	7·48	0·0404M	0·0576M	1·42	0·152
15·25	7·06	0·0382M	0·0554M	1·45	0·161
30·45	6·65	0·0360M	0·0532M	1·48	0·170
45·25	6·30	0·0341M	0·0513M	1·51	0·179
59·35	6·00	0·0324M	0·0496M	1·53	0·185
73·45	5·71	0·0309M	0·0481M	1·56	0·193
90·15	5·45	0·0295M	0·0467M	1·58	0·199
104·40	5·24	0·0283M	0·0455M	1·61	0·207
119·25	5·03	0·0272M	0·0444M	1·63	0·213
134·45	4·83	0·0262M	0·0434M	1·66	0·220

Plotting of $\log_{10} c^B/c^A$ ($\log_{10} [MeI]/[EtO^\ominus]$) against t results in a straight line (Fig. 1.7):

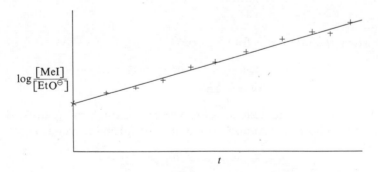

Fig. 1.7

Measurement of its slope, which is equal to $([MeI]_0 - [EtO^\ominus]_0)/2\cdot303k$, thus gives a value of k (at 21°) of $6\cdot59 \times 10^{-2}$; its units are litres moles^{-1} min^{-1}.

(c) Other order reactions

Thus for the essentially irreversible reaction

$$A + B + C + \cdots \longrightarrow D + E + F + \cdots$$

a generalized rate equation can be written of the form:

$$\text{Rate} = k[A]^x[B]^y[C]^{z\cdots}$$

The powers x, y and z can have values not only of 1 and 2, as we have seen above, but also 0 (zero order), -1 (i.e. the rate of reaction is $\propto$ the reciprocal of, for example $[A]$), and even non-integral values.

Overall third order reactions are relatively common, though considerably less so than overall second order reactions. Thus for the Mannich reaction of acetophenone (19)

$$PhCOCH_3 + CH_2O + HNMe_2 \longrightarrow PhCOCH_2CH_2NMe_2 + H_2O$$
(19)

the rate law is found to be

$$\text{Rate} = k[PhCOCH_3][CH_2O][HNMe_2]$$

i.e. it is first order in each reactant, so that $x = y = z = 1$. Overall third order reactions in A and B only are probably more often met with, however, as in the benzoin condensation of benzaldehyde (20),

$$2PhCHO \xrightarrow{\ominus CN} PhCH(OH)COPh$$
(20)

for which the rate law is found to be

$$\text{Rate} = k[PhCHO]^2[\ominus CN]$$

i.e. $x = 2$, $y = 1$

A few fourth order reactions are known—though rarely in A, B, C and D—such as the Cannizzaro reaction of formaldehyde (21),

$$2CH_2O + \ominus OH \longrightarrow HCO_2{}^{\ominus} + CH_3OH$$
(21)

for which, in very high concentrations of base, the rate law is found to be:

$$\text{Rate} = k[CH_2O]^2[\ominus OH]^2$$

i.e. $x = y = 2$

A reaction that is zero order in respect of a species appearing in the stoichiometric equation is the acid-catalysed iodination of acetone (22) (*cf.* pp. 2, 38),

$$CH_3COCH_3 + I_2 \xrightarrow{H^\oplus} CH_3COCH_2I + H^\oplus I^\ominus$$
(22)

for which the rate law is found to be,

$$Rate = k[CH_3COCH_3][H^\oplus]$$

i.e. $x = y = 1, z = 0$

i.e. the reaction is overall second order, but it is zero order in respect of iodine though this must obviously be involved at some stage in the overall reaction (*cf.* p. 38) as it becomes incorporated in the product! Non-integral values of x, y and z are also sometimes observed as in the benzidine rearrangement (*cf.* pp. 59, 216) of the N,N'-diaryl-hydrazine (23) in 95 per cent aqueous ethanol

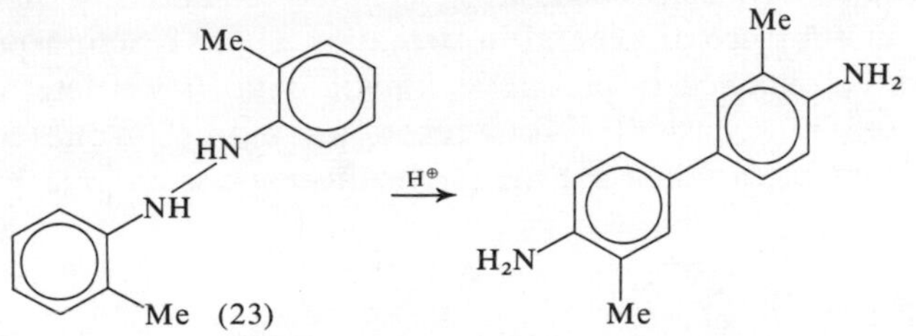

for which the rate law is found to be:

$$Rate = k[ArNHNHAr][H^\oplus]^{1 \cdot 6}$$

i.e. $x = 1, y = 1 \cdot 6$

In this case it can, however, be shown that the non-integral order results from the simultaneous operation of alternative reaction pathways that are first and second order, respectively, in $[H^\oplus]$. The general overall rate law is thus of the form,

$$Rate = k_1[ArNHNHAr][H^\oplus] + k_2[ArNHNHAr][H^\oplus]^2$$

i.e. $x = y = 1, and\ x = 1, y = 2$

and the observed non-integral order ($y = 1 \cdot 6$) is an averaged one reflecting the proportion of the total reaction that is proceeding by each mode. Non-integral values are also observed in a number of reactions that involve radicals as intermediates (*chain reactions, cf.* p. 29).

An example is the thermal decomposition of acetaldehyde (24)

$$CH_3CHO \xrightarrow{\Delta} CH_4 + CO$$
(24)

for which the rate law is found to be:

$$Rate = k[CH_3CHO]^{1.5}$$

i.e. $x = 1.5$.

Finally, it must be emphasized that reactions are known for which the kinetic data cannot be accommodated by a simple generalized rate equation such as that above: the concept of reaction order then no longer applies.

RATE CONSTANTS AND ACTIVATION ENERGY

Before proceeding to consider the vital question of the interpretation of experimentally determined rate laws in molecular terms, there is one further useful piece of general information about reaction pathways that can be obtained from simple kinetic data. If we determine the values of the rate constant, k, for a reaction at more than one (constant) temperature we can then use the Arrhenius expression [9],

$$k = Ae^{-E/RT}$$
[9]

relating the rate constant to the absolute temperature T, in order to calculate E, the activation energy of the reaction (*cf.* p. 8). In the above expression R is the gas constant (8.32 Joules mole^{-1}deg$^{-1} \equiv 1.99$ cal mole^{-1}deg^{-1}), and A a constant for the reaction—independent of temperature—that is related to the proportion of the total number of collisions between reactant molecules that result in the successful formation of products. The derivation of the above equation, [9], is somewhat shaky on theoretical grounds, at least so far as giving wholly adequate physical justification to A and E is concerned, but it has proved extremely convenient in the semi-empirical ordering of experimental results so as to yield values of E that are most useful for comparative purposes.

The relation is more readily usable in the form [10];

$$\log_{10} k = -\frac{E}{2.303\,RT} + \log_{10} A$$
[10]

then by determining, experimentally, the value of the rate constant at two different temperatures, k_1 at temperature T_1 and k_2 at T_2, and subtracting $\log_{10}k_2$ from $\log_{10}k_1$ we obtain the equation [11],

$$\log_{10}\frac{k_1}{k_2} = -\frac{E}{2\cdot303\,R}\left[\frac{1}{T_1}-\frac{1}{T_2}\right] \qquad [11]$$

whence the value of E may be calculated. Alternatively, and generally more satisfactorily, we can determine the values of k for two or more different temperatures and then plot $\log_{10}k$ against $1/T$ (*cf.* Fig. 1.8). We should, from [10], get a straight line whose slope is $-E/2\cdot303\,R$; the slope may be measured, and a value of E thereby obtained. Thus in the solvolysis of Me_3CCl that we have already referred to (p. 14), experimentally determined 10^3k values at 288°, 298° and 307°K were found to be 1·09, 3·06, and 9·48, respectively. These result in the plot below (Fig. 1.8):

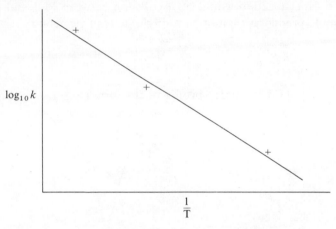

Fig. 1.8

Measurement of its slope, and subsequent calculation, yields an E value of 88 kJ/mole (21 kcal/mole) for the solvolysis.

As has already been mentioned (p. 8) the activation energy, E, may be roughly equated with ΔH^+, the enthalpy of activation, and comprises the energy necessary to begin the stretching of, and increase the vibration in, bonds that have to be broken in the overall reaction.

THE INTERPRETATION OF RATE EQUATIONS IN MOLECULAR TERMS

Now that we have seen how it is possible to establish a rate equation for a reaction from the results of kinetic experiments, it remains to find out exactly what such a rate equation can be made to tell us about the specific molecular interactions through whose agency starting materials are converted into products.

(a) Second order

As we saw above (p. 17) the experimentally determined rate law for the reaction of iodomethane (25) with ethoxide ion in ethanol

$$EtO^{\ominus} + MeI \longrightarrow EtOMe + I^{\ominus}$$
$$(25)$$

was found to be:

$$Rate = k[MeI][EtO^{\ominus}]$$

The simplest possible interpretation in molecular terms of this overall second order reaction would be a collision between MeI and $EtO^{\ominus}$, leading to product formation in a single step (an elementary reaction),

$$EtO^{\ominus} + Me-I \longrightarrow \overset{\delta-}{EtO}\cdots Me\cdots \overset{\delta-}{I} \longrightarrow EtO-Me + I^{\ominus}$$
$$(25)$$

corresponding to an energy profile of the form (Fig. 1.9, *cf.* Fig. 1.3, p. 7):

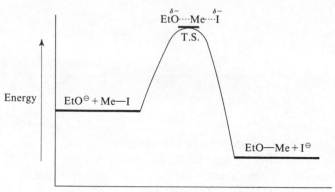

Fig. 1.9

A pathway of this nature is generally referred to as a *bimolecular* reaction; the molecularity of a reaction may be defined as the number of species (molecules, ions etc.) that are necessarily undergoing covalency change in its rate-limiting step (*cf.* p. 2). In the above example, an elementary reaction, the rate-limiting step is synonymous with the overall reaction, and two species are undergoing covalency changes during its course. This particular pathway for the above displacement reaction is referred to as S_N2—substitution nucleophilic bimolecular. Such a concerted pathway for the reaction would have the manifest advantage that the new (EtO—Me) bond is beginning to form, and is thereby providing energy, as soon as the old (Me—I) bond begins to break; such an energy 'down-payment' so early in the reaction is likely to mean that the $\Delta H^{\ddagger}$ component ($\approx E$, the activation energy) of $\Delta G^{\ddagger}$ may well be relatively small, and the reaction correspondingly rapid.

A clear distinction needs to be made between the order of a reaction and its molecularity: the former is an experimentally determined fact about a reaction, while the latter has significance only in the light of a particular pathway for the reaction, i.e. it is part of the mechanistic interpretation of the reaction and is susceptible to re-evaluation, in the light of additional information, in a way that the order cannot be. In the above elementary reaction the two are synonymous in that the reaction is both of the second order and bimolecular, but such coincidence is not universal as we shall see below (p. 29).

(b) First order

As was seen previously (p. 14) the experimentally determined rate law for the solvolysis of the tertiary halide 2-chloro-2-methylpropane (26) in aqueous acetone

$$H_2O: + (CH_3)_3C-Cl \longrightarrow HO-C(CH_3)_3 + H^{\oplus}Cl^{\ominus}$$
$$(26) \qquad\qquad\qquad (27)$$

was found to be:

$$\text{Rate} = k[(CH_3)_3C-Cl]$$

The reaction is thus first order overall and, on the basis of the above rate law, the slow, rate-limiting step of the reaction apparently involves the halide (26) only. As water is necessarily participant at some point in the overall reaction—only it can provide the oxygen atom in the product—the solvolysis must proceed through at least one further stage: that in which oxygen incorporation takes place. The simplest possible interpretation of this first order reaction would thus have the general form,

$$(CH_3)_3C-Cl \xrightarrow[\text{(i) slow}]{} \text{intermediate} \xrightarrow[\text{(ii) fast}]{H_2O:} (CH_3)_3C-OH + H^{\oplus}Cl^{\ominus}$$
$$(26) \qquad\qquad\qquad\qquad\qquad\qquad (27)$$

and a not unreasonable candidate for the intermediate might be the *ion-pair* $(CH_3)_3C^{\oplus}Cl^{\ominus}$ (28). The slow, rate-limiting formation of this ion pair (*cf.* p. 128) must involve the breaking of the C—Cl bond, i.e. a furthering of $(CH_3)_3C^{\underline{\delta+}}Cl^{\delta-}$, but this will be assisted by the energy-yielding possibilities of solvation of the developing ion pair by the highly polar and—more important—hydroxylic solvent. This is then followed by a further, fast, non-rate limiting stage in which H_2O: attacks the highly reactive carbonium ion, $(CH_3)_3C^{\oplus}$, of the ion pair (28) to yield products:

$$(CH_3)_3C-Cl \xrightarrow[\text{(i) slow}]{} (CH_3)_3C^{\oplus}Cl^{\ominus} \xrightarrow[\text{(ii) fast}]{H_2O:} (CH_3)_3C-OH + H^{\oplus}Cl^{\ominus}$$
$$\qquad(26)\qquad\qquad\qquad(28)\qquad\qquad\qquad(27)$$

Support for the intermediate being something like (28) is provided by the fact that carrying out the solvolysis in aqueous ethanol, rather than aqueous acetone, leads to the formation of the ethyl ether (29) in addition to the alcohol (27). Even more cogent evidence is the formation, in either solvent, of not inconsiderable amounts of the alkene (30), that could be formed by the loss of a proton from the carbonium ion of (28):

$$
\begin{array}{ll}
 & \xrightarrow{H_2O:} (CH_3)_3C-OH + H^{\oplus}Cl^{\ominus} \\
 & \qquad\qquad\quad (27) \\
(CH_3)_3C^{\oplus} \xrightarrow{Et\ddot{O}H} & (CH_3)_3C-OEt + H^{\oplus}Cl^{\ominus} \\
\quad\; Cl^{\ominus} & \qquad\qquad\quad (29) \\
(28) \xrightarrow{-H^{\oplus}} & (CH_3)_2C{=}CH_2 + H^{\oplus}Cl^{\ominus} \\
 & \qquad\qquad\quad (30)
\end{array}
$$

The two-step reaction pathway above corresponds to an energy profile of the form (Fig. 1.10):

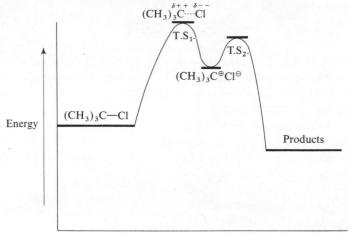

Fig. 1.10

The ion-pair intermediate (28) constitutes an energy *minimum* in the reaction profile between the overall energy maximum (T.S$_1$.) for the slow rate-limiting step (in this case the first one), and the smaller energy maximum (T.S$_2$.) for the fast (less energy-demanding) second step in which the ion-pair intermediate (28) is converted into products. This particular pathway for the solvolysis is referred to as S$_N$1—substitution nucleophilic unimolecular.

It is important to realise that the data obtained from kinetic measurements establish basic criteria that any suggested reaction pathway must fulfil if it is to merit any further serious consideration: they do not, and indeed cannot, *prove* that a reaction proceeds by a specific pathway. Thus an alternative route for the above solvolysis could be:

$$H_2O\!:\,+\,(CH_3)_3C\!-\!Cl \xrightarrow[\text{(i) slow}]{} \overset{\oplus}{HO}\!-\!C(CH_3)_3 \xrightarrow[\text{(ii) fast}]{} HO\!-\!C(CH_3)_3 + H^{\oplus}Cl^{\ominus}$$
$$(26) \qquad\qquad\quad \underset{+\,Cl^{\ominus}}{H}$$

This would, as written, be expected to lead to a kinetic rate law of the form,

$$\text{Rate} = k[(CH_3)_3C\!-\!Cl]\,[H_2O\!:]$$

but the water is present, as solvent, in such large excess that even if it did participate as shown above its concentration would be essentially the same at the end of the reaction as it was at the beginning—it would change by <0.2 per cent during the complete solvolysis of a 0.1 molar solution of the halide. The projected rate law for the bimolecular reaction above would thus become,

$$\text{Rate} = k_{\text{obs.}}[(CH_3)_3C\!-\!Cl]$$

i.e. this is what would actually be obtained from kinetic measurements, and it is not possible to tell therefore, from such simple kinetic measurements, by which pathway the solvolysis actually proceeds. That the bimolecular pathway is much less likely than the unimolecular is shown by the fact that the reaction rate is not changed by the addition of base (*cf.* p. 14). We would expect $^{\ominus}OH$ to be a very much better nucleophile than H_2O: and the fact that the former is not involved in the rate-limiting step of the solvolysis makes it most unlikely that the latter can be. Further light may be thrown on the detailed pathway of such solvolyses—and their unimolecular nature thereby more firmly established—by a study of their stereochemistry (*cf.* p. 128).

(c) Order and the composition of the transition state

It may be seen from the above examples that the major information, in molecular terms, that the rate law gives us is that it establishes the *composition* of the transition state. It tells us which of the original reactants go to make up the transition state, and in what relative proportions, either for the single step in an elementary reaction (e.g. $EtO^{\ominus}$ + MeI above, p. 22), or for the slow, rate-limiting step in a complex reaction (e.g. $(CH_3)_3C—Cl + H_2O$: above, p. 23).

It thus follows for the benzoin condensation of benzaldehyde (31)

$$2PhCHO \xrightarrow{\ominus CN} PhCH(OH)COPh$$
$$(31) \qquad\qquad (36)$$

that because the experimentally determined rate law is found to be,

$$Rate = k[PhCHO]^2[^{\ominus}CN]$$

then the transition state of this third order reaction must involve two molecules of benzaldehyde and one cyanide ion. It does *not* however follow that the reaction proceeds through a simple three-body collision between these species: such termolecular reactions are indeed extremely rare, not least because the entropy of activation (p. 8) would be expected to be very high. Equally compatible with the observed rate law—and statistically much more likely—would be two or more successive steps to reach the transition state, the first involving two of the three species and the resultant adduct then going on, in a further step, to react with the third: the only requirement is that the first step must be fast and the second slow, or the reaction would have a second order rate law involving only the two reactants involved in the first step.

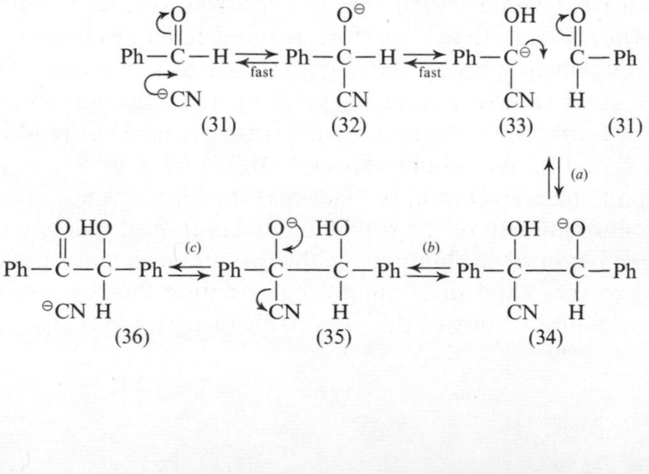

The third order benzoin condensation is believed to follow the pathway, in which necessarily rapid attack of $^\ominus$CN on PhCHO (31) yields the adduct (32), and this is followed by necessarily rapid proton transfer to form the carbanion (33). This can then act as a nucleophile and attack a second molecule of benzaldehyde to yield the adduct (34), proton transfer then yields (35) which can eliminate cyanide ion to form the product benzoin (36). Any of the stages (*a*), (*b*) or (*c*) could be rate-limiting and the overall reaction would still follow the observed third order rate law; however, proton transfers, such as (*b*), between electronegative atoms are very often fast, and (*c*) is the reversal of cyanohydrin formation which is also likely to be fairly rapid under these conditions. It would thus seem most likely that (*a*) is here the slow, rate-limiting stage of the reaction, and this is certainly compatible with its involving the formation of a carbon–carbon bond, which is very often a relatively slow process.

An important point emerges from this simple analysis: though the experimentally determined rate law tells us the *composition* of the transition state, it does not—except by indirect inference—tell us anything about its *structure*. The full, and useful, interpretation of rate laws in molecular terms requires the utilization of our general chemical knowledge and experience, and the exercise of a creative imagination, in order to make reasonable suggestions about the structure of the transition state, and to decide which steps involved in its formation and decomposition are fast and which are slow. The reaction scheme suggested above would thus require the transition state—corresponding to the energy maximum in the overall reaction—to have a structure something like (37):

$$
\begin{array}{ccc}
\text{OH} & & \text{O}^{\delta-} \\
|_{\delta-} & & \| \\
\text{Ph}-\text{C}\cdots\cdots\cdots\text{C}-\text{Ph} \\
| & & | \\
\text{C}\equiv\text{N} & & \text{H}
\end{array}
$$
(37)

The benzoin condensation has been shown to be reversible overall, and a further point of interest about it is the very highly specific catalysis by $^\ominus$CN: it is not catalysed by $^\ominus$OH, nor by other $Y^\ominus$ species. It seems likely that apart from its ready addition to (first step), and elimination from (step *c*), C=O—which could be duplicated by other $Y^\ominus$ species— the really significant role of $^\ominus$CN lies in its ability to stabilize the carbanion (33), and the minor transition state that precedes it, by delocalization,

$$
\begin{array}{ccccc}
\text{O}^\ominus & & \text{OH} & & \text{OH} \\
| & & | & & | \\
\text{Ph}-\text{C}-\text{H} \longrightarrow & \left[\; \text{Ph}-\text{C}^\ominus \right. & \longleftrightarrow & \left. \text{Ph}-\text{C} \;\right] \\
| & & | & & \| \\
\text{C}\equiv\text{N} & & \text{C}\equiv\text{N} & & \text{C}=\text{N}^\ominus \\
(32) & & (33a) & & (33b)
\end{array}
$$

thus ensuring its rapid formation from (32).

Examples such as the above in which there are rapid steps, often equilibria, before the rate-limiting stage of a higher than second order reaction are very common; one is analysed in more detail below (p. 217).

(d) Zero order terms—the iodination of acetone

We have already referred (p. 2) to the acid-catalysed iodination of acetone (38)

$$CH_3COCH_3 + I_2 \xrightarrow{H^\oplus} CH_3COCH_2I + H^\oplus I^\ominus$$
$$(38)$$

for which—somewhat unexpectedly—the rate law was found to be,

$$Rate = k[CH_3COCH_3][H^\oplus]$$

i.e. the reaction is zero order with respect to iodine. The slow, rate-limiting step of the iodination thus involves acetone and the acid catalyst only, and the iodine that is ultimately incorporated into the product must become so in a fast step *beyond* the rate-limiting one:

$$CH_3COCH_3 + H^\oplus \xrightarrow[\text{(i) slow}]{} \text{intermediate} \xrightarrow[\text{(ii) fast}]{I_2} CH_3COCH_2I + H^\oplus I^\ominus$$
$$(38)$$

It comes as no surprise to find acetone (38) undergoing protonation, for this is known to happen readily, and reversibly, on the oxygen atom of carbonyl groups:

$$CH_3-\overset{\overset{\textstyle O}{\|}}{C}-CH_3 + HY \rightleftharpoons \left[CH_3-\overset{\overset{\textstyle \oplus}{\overset{\textstyle OH}{\|}}}{C}-CH_3 \leftrightarrow CH_3-\overset{\overset{\textstyle OH}{|}}{\underset{\oplus}{C}}-CH_3 \right] + Y^\ominus$$
$$(38) \qquad\qquad\qquad\qquad\qquad\qquad\qquad (39)$$

The resultant carbonium ion (39) does not, however, look like a very convincing intermediate for the iodination on the grounds (*a*) that there seems no reason why iodine should attack it especially readily in a fast step, and (*b*) protonations on oxygen, such as that which led to its formation, are generally fast and not slow as (i) above requires. It was therefore suggested that the significant intermediate might be obtained by slow loss of proton from *carbon* in (39) to yield the enol form of acetone (40), whose activated carbon–carbon double bond might indeed be expected to react readily with iodine in a fast step. The possible overall reaction pathway would thus be:

$$CH_3-\overset{\overset{\displaystyle O}{\|}}{C}-CH_3 \underset{\text{fast}}{\overset{H_3O^\oplus}{\rightleftarrows}} CH_3-\overset{\overset{\displaystyle OH}{|}}{\underset{\oplus}{C}}-CH_2 \underset{\text{slow}}{\overset{-H_3O^\oplus}{\rightleftarrows}} CH_3-\overset{\overset{\displaystyle O-H}{|}}{C}=CH_2 \underset{\text{fast}}{\longrightarrow} CH_3-\overset{\overset{\displaystyle O}{\|}}{C}-CH_2 + H^\oplus I^\ominus$$

(38)   (39)   (40)

The experimentally determined rate law for a reaction thus tells us nothing about any steps beyond its rate-limiting stage except that they must be fast; we may, however, be able to speculate about the species involved in them, particularly if one or more of these appear in the stoichiometry of the reaction though not in its rate law—as in the case above.

It should be noted that acid is formed as one of the products in the above acid-catalysed iodination, and the concentration of catalyst thus rises progressively as the reaction proceeds. It is thus said to be *autocatalysed*, and a progressively increasing reaction rate will be observed provided the initial concentration of acid is not too large. Once again, additional evidence in favour of the above type of reaction pathway is provided by stereochemical studies (*cf.* p. 164).

(e) Non-integral orders. (i) The pyrolysis of acetaldehyde

In the light of what has been said above it is clearly possible to translate rate equations involving integral powers of some, though not necessarily all, of the stoichiometric reactants into meaningful representations of reaction pathways in molecular terms. Such a translation does not, however, spring instantly to mind for a reaction such as the thermal decomposition of acetaldehyde (41),

$$CH_3CHO \overset{\Delta}{\longrightarrow} CH_4 + CO$$
(41)

for which the experimentally determined rate-law is found to be (*cf.* p. 20):

$$\text{Rate} = k_{\text{obs.}}[CH_3CHO]^{3/2}$$

Clearly the rate-limiting stage of the reaction cannot involve half a molecule of acetaldehyde! A more plausible suggestion that has been made is that the reaction proceeds through radical intermediates in a so-called *chain reaction pathway* involving the following steps:

Initiation: $CH_3CHO \xrightarrow{k_1} CH_3 \cdot + \cdot CHO$ (a)
                 (41)

Propagation: $CH_3 \cdot + CH_3CHO \xrightarrow{k_2} CH_4 + CO + CH_3 \cdot$ (b)
                  (41)

Termination: $2CH_3 \cdot \xrightarrow{k_3} CH_3CH_3$ (c)

                  $2 \cdot CHO \xrightarrow{k_4} 2CO + H_2$ (d)

When the reaction is proceeding at a steady rate (the so-called steady state assumption) we can assume that the rate of formation of $CH_3 \cdot$ in (a) ($= k_1[CH_3CHO]$) is equal to the rate of its destruction in (c) ($= k_3[CH_3 \cdot]^2$), this must be so as (b) is without effect on $[CH_3 \cdot]$. Thus

$$k_3[CH_3 \cdot]^2 = k_1[CH_3CHO]$$

whence: $[CH_3 \cdot] = \left(\dfrac{k_1}{k_3}[CH_3CHO]\right)^{1/2}$ [12]

The rate of formation of products from starting materials, i.e. the overall rate, is determined by (b), hence

$$\text{Rate} = k_2[CH_3 \cdot][CH_3CHO]$$

and substitution of $[CH_3 \cdot]$ from [12] above leads to:

$$\text{Rate} = k_2 \left(\frac{k_1}{k_3}\right)^{1/2} [CH_3CHO]^{3/2} = k_{obs.}[CH_3CHO]^{3/2}$$

In general support of such a pathway, radicals can be detected in the above reaction (e.g., by the etching of metals mirrors, e.s.r spectroscopy (*cf.* p. 88) etc.), it is inhibited by the usual radical inhibitors (*cf.* p. 74), and some ethane and hydrogen (from (c) and (d), respectively) can be detected in the products.

(ii) Competing pathways:

Non-integral orders of reaction with respect to a particular reactant may also be observed when starting materials can be converted into products by alternative reaction pathways operating simultaneously. In this connection, we saw above that nucleophilic displacement reactions on methyl halides followed a bimolecular pathway (p. 22), while similar attack on *t*-butyl halides followed a unimolecular mode. It thus follows that somewhere along the series

$$CH_3-X \quad MeCH_2-X \quad Me_2CH-X \quad Me_3C-X$$

$$(X = Hal)$$

a changeover of mechanistic pathway must take place, and it is significant that the base-induced hydrolysis of Me_2CH-Br is found to follow a rate law of the form,

$$\text{Rate} = k[Me_2CH-Br][^{\ominus}OH]^n$$

where n lies between 0 and 1 depending on the initial concentration of hydroxyl ion: the greater the initial $[^{\ominus}OH]$, the more closely n is found to approach 1.

In the light of the mechanistic changeover in the series above, it seems most likely that the hydrolysis of Me_2CHBr is proceeding not by a special, unique pathway, but by the simultaneous operation of uni- and bi-molecular modes. This would be reflected in a general rate law of the form,

$$\text{Rate} = k_1[Me_2CH-Br] + k_2[Me_2CH-Br][^{\ominus}OH]$$

the relative proportions of the total reaction proceeding by each mode defining n in the experimentally observed rate law above. We should expect the contribution of the second (k_2) term to increase as initial $[^{\ominus}OH]$ is increased, as is indeed observed in practice. In a case such as this, and also that of the benzidine rearrangement referred to above (p. 19), it is often possible to analyse the raw kinetic data so as to evaluate k_1 and k_2 and to assess the relative contributions, under a given set of conditions, of the competing pathways.

(f) The medium and the nature of the transition state

It is found in practice that the rates—though not necessarily the rate equations—of many reactions are influenced by the addition of neutral salts. This is known as a *salt effect* and results from the influence of the resultant increased ionic strength of the medium on a step or steps of a reaction in which charge is concentrated or dispersed; this change in charge distribution may be made energetically less or more difficult, as the case may be, with corresponding effects on the rates of the steps involved. Broadly speaking, a positive salt effect—one which results in an increase of reaction rate—points to a transition state, for the rate-limiting step of the overall reaction, that is more polar than the species which preceded it, which may, of course, be either the starting materials or an intermediate. This can be of considerable diagnostic value when considering potential models for a transition state (*cf.* p. 219).

Change of solvent with consequent change in polarity, which embraces both dielectric constant and potential ion-solvating ability, can also have a profound effect on the rate of a particular reaction (*cf.* p. 193). Here too, increasing reaction rate with increasing solvent polarity can be used diagnostically to recognize a transition state that is more polar than the species that preceded it.

REVERSIBLE REACTIONS

In virtually all the examples we have considered to date it has been assumed that, for kinetic purposes, the conversion of reactants into products is essentially irreversible. This is indeed true for many organic reactions but not of course for all of them, and for the reversible elementary reaction,

$$A + B \underset{k_r}{\overset{k_f}{\rightleftharpoons}} C + D$$

the rate will be determined by:

$$\text{Rate} = \frac{-d[A]}{dt} = k_f[A][B] - k_r[C][D]$$

If kinetic measurements are made in the early stages of such a reaction, it may well be that the second term—the one due to the competing, reverse reaction—is negligible, but this may remain true for so short a time (i.e. for so small a fraction of the reaction's total course) as not to allow of an unequivocal determination of the forward reaction's order.

When the position of equilibrium is reached there is no further nett change in the concentration of A; the equilibrium is, of course, a dynamic one in that A + B are still being converted into C + D and *vice versa*, but there is no further *nett* change in the concentration of any of the species involved. $-d[A]/dt$ thus becomes equal to zero:

$$O = k_f[A][B] - k_r[C][D]$$
$$k_r[C][D] = k_f[A][B]$$
$$\frac{[C][D]}{[A][B]} = \frac{k_f}{k_r}$$

The last expression above is, by definition, also equal to K, the classical equilibrium constant for the reaction; thus $K = k_f/k_r$ and, if k_f and K can be evaluated experimentally, k_r may then be obtained without further experiment.

It cannot be too strongly emphasized that simple considerations such as the above apply only to elementary reactions, that authentic elementary reactions are comparatively rare, and that they may well be somewhat difficult to identify experimentally in any case. When we come to consider reversible reactions that are complex (i.e. multi-step), the rate expressions we should need for K to be equal to k_f/k_r would be those that apply *at equilibrium*; our actual kinetic measurements are, however, made under non-equilibrium conditions, and lead therefore to non-equilibrium rate expressions which may well differ markedly from the 'theoretical' equilibrium ones. It is, of course, possible—provided one is prepared to go to the trouble—to measure reaction rates actually at equilibrium by the use of isotopic tracers. Thus in the hydrolysis of ethyl acetate (42),

$$CH_3CO_2C_2H_5 + H_2O \underset{k_r^{eq.}}{\overset{k_f^{eq.}}{\rightleftharpoons}} CH_3CO_2H + C_2H_5OH$$
$$(42)$$

$k_f^{eq.}$ may be evaluated directly by making up the above four-constituent mixture at the known equilibrium concentrations, but using ethyl acetate that has been enriched in the radioactive isotope ^{14}C (cf. p. 58) in its methyl group (42a), $^{14}CH_3CO_2C_2H_5$, and observing the rate at which ^{14}C appears in the acetic acid:

$$^{14}CH_3CO_2C_2H_5 + H_2O \xrightarrow{k_f^{eq.}} {}^{14}CH_3CO_2H + C_2H_5OH$$
$$(42a)$$

The unfortunately still common practice of equating K with k_f/k_r (i.e. not $k_f^{eq.}$ and $k_r^{eq.}$) for *all* reactions is much to be deplored, for the contribution of one particular step to the overall rate of reaction may well change considerably as the reaction proceeds towards the attainment of equilibrium.

One further point needs to be made here about reversible reactions: that under the same conditions, the reaction pathway, energy profile etc., of the backward reaction must be the exact reverse of those for the forward reaction; this is known as the *principle of microscopic reversibility*. Because the same intermediates and transition states are traversed, albeit in the reverse sequence, the principle can be of use in suggesting structures for such species, and in identifying possible fast and slow steps etc., as some of the steps may be subject to more ready interpretation by our chemical insight in the backward, than in the forward, direction.

KINETIC *v*. THERMODYNAMIC CONTROL OF PRODUCT COMPOSITION

In a reaction involving the formation of alternative products from the same starting materials—a fairly common occurrence in organic reactions—we might expect the relative proportions of the alternative products obtained to be related to their respective rates of formation: the faster the rate at which a particular product is formed, the more of it there will be in the total product mixture. Thus in the nitration of chlorobenzene (43),

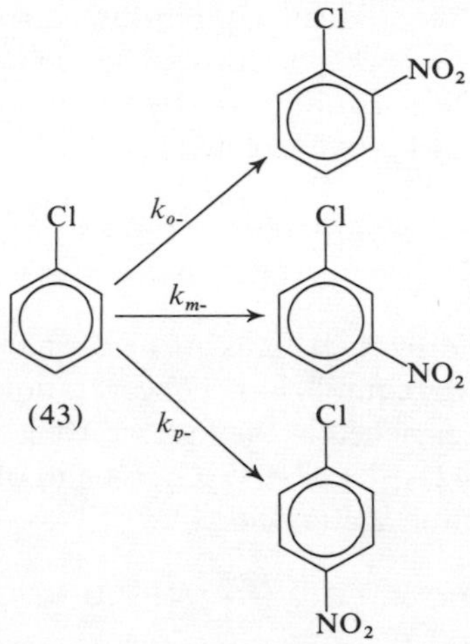

it may be shown that, under specified conditions, the formation of each of the three (*o*-, *m*- and *p*-) isomeric nitrochlorobenzenes is first order with respect to chlorobenzene,

$$\frac{d[o\text{-}ClC_6H_4NO_2]}{dt} = k_{o\text{-}}[C_6H_5Cl],$$

$$\frac{d[m\text{-}ClC_6H_4NO_2]}{dt} = k_{m\text{-}}[C_6H_5Cl],$$

$$\frac{d[p\text{-}ClC_6H_4NO_2]}{dt} = k_{p\text{-}}[C_6H_5Cl]$$

and that the disappearance of chlorobenzene is also first order with respect to chlorobenzene:

$$\frac{-d[C_6H_5Cl]}{dt} = k[C_6H_5Cl]$$

where $k = k_{o\text{-}} + k_{m\text{-}} + k_{p\text{-}}$

It may thus be shown, from the differential equations for product formation, that:

$$[o\text{-ClC}_6\text{H}_4\text{NO}_2] : [m\text{-ClC}_6\text{H}_4\text{NO}_2] : [p\text{-ClC}_6\text{H}_4\text{NO}_2] = k_{o\text{-}} : k_{m\text{-}} : k_{p\text{-}}$$

Determination of the product ratios (generally by a physical method, e.g. spectroscopy, rather than by isolation) after any given time interval thus enables us to evaluate the rate constants. This holds generally, irrespective of the order of the reactions by which the products are formed, provided only that the order is the same in each case. Such an overall situation is generally referred to as *kinetic control* of product composition.

In some cases, however, the product ratio is found not to be constant but to change as the reaction proceeds; thus in the Friedel–Crafts alkylation of toluene (44) with benzyl bromide (GaBr$_3$ as the Lewis acid catalyst) at 25°,

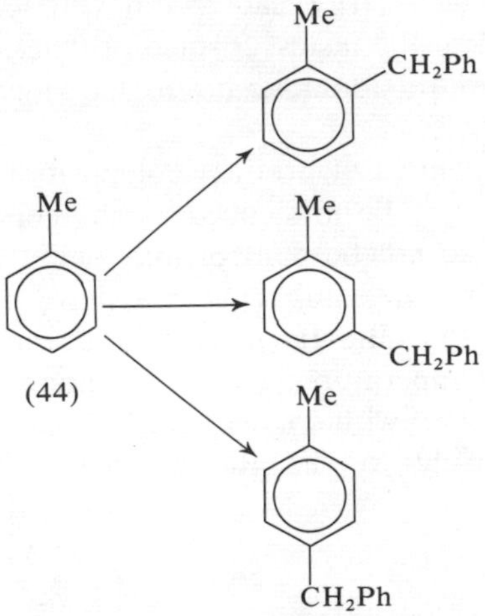

the isomer distributions in the total substitution product obtained to date were, after two successive time intervals, found to be:

sec	%o-	%m-	%p-
0·01	40	21	39
10	23	46	31

Clearly simple kinetic control of product composition cannot be operative for this would lead to a constant isomer distribution, and it is significant that the distribution after ten seconds closely reflects that to be expected from the known, relative thermodynamic stabilities of the three isomers. The product composition is now being controlled, not by the relative rates of formation of the three alternative isomeric products (kinetic control), but by their relative thermodynamic stabilities under the conditions of the reaction: such an overall situation is generally referred to as *thermodynamic control* of product composition.

It can readily be shown that any of the above three benzyltoluenes can be converted into a mixture of all three, under the conditions of the original Friedel–Crafts reaction. The conditions are thus such as to allow of an equilibrium to be set up between the three products (either directly between each other, or *via* dealkylation to toluene followed by realkylation to produce a different isomer), so no matter which is formed the fastest, ultimately it will be the most stable, under the conditions of the reaction, that will predominate; this situation is thus also referred to as *equilibrium control* of product composition. The two modes of product composition control, kinetic and thermodynamic, need not necessarily lead to different sequences of relative product abundance: if the fastest formed product also happens to be the most stable it will predominate no matter which of the two modes of control is operative.

Which mode of control of product composition is operative for a particular reaction may be influenced by the conditions. Thus over short periods of time kinetic control may operate, but after longer exposure a relatively slow equilibrium may have had time to become established so that the ultimate product is thermodynamically controlled. Changes in temperature may also be important; thus sulphonation of naphthalene (45) with concentrated sulphuric acid at 80° leads virtually quantitatively to the thermodynamically less stable 1-sulphonic acid (46),

(45)   (46)

and kinetic control must thus be operative. Similar sulphonation at 160° leads largely to the thermodynamically more stable 2-sulphonic acid (47),

(45)   (47)   81%   (46)   19%

and there must thus have been a shift, with the rise of temperature, to thermodynamic control. This particular reaction is considered somewhat further below (p. 57).

FURTHER READING

ATHERTON, M. A. and LAWRENCE, J. K. *An Experimental Introduction to Reaction Kinetics* (Longman, 1970).

BAMFORD, C. H. and TIPPER, C. F. H. (Eds.) *Comprehensive Chemical Kinetics, Vol. I: The Practice of Kinetics* (Elsevier, 1969).

BENSON, S. W. *Foundations of Chemical Kinetics* (McGraw Hill, 1960).

FRIESS, S. L., LEWIS, E. S. and WEISSBERGER, A. (Eds.) *Technique of Organic Chemistry, Vol. VIII, Pts. I and II: Investigation of Rates and Mechanisms of Reactions* (Interscience, 1961, 1963).

FROST, A. A. and PEARSON, R. G. *Kinetics and Mechanism* (Wiley, 2nd Ed. 1961).

GARDINER, W. C. *Rates and Mechanisms of Chemical Reactions* (Benjamin, 1969).

GOULD, E. S. *Mechanism and Structure in Organic Chemistry* (Holt, Rinehart and Winston, 1959), pp. 159–194.

HAMMETT, L. P. *Physical Organic Chemistry* (McGraw Hill, 2nd Ed. 1970) pp. 53–145.

LAIDLER, K. J. *Reaction Kinetics, Vol. 2: Reactions in Solution* (Pergamon, 1963).

LATHAM, J. L. *Elementary Reaction Kinetics* (Butterworth, 2nd Ed. 1969).

LEFFLER, J. E. and GRUNWALD, E. *Rates and Equilibria of Organic Reactions* (Wiley, 1963), pp. 57–127.

MENZINGER, M. and WOLFGANG, R. 'The Meaning and Use of the Arrhenius Activation Energy', *Angew. Chem. Int. Ed.,* 1969, **8**, 438.

2
The uses of isotopes—kinetic and non-kinetic

We have seen in the last chapter the quantity and quality of information about reaction pathways that kinetic measurements are able to provide us with; sophisticated as this is there comes a point at which they can, hardly surprisingly, offer no further assistance. Thus simple kinetic data cannot normally be made to tell us whether a particular bond has, or has not, been broken in a step up to and including the rate-limiting stage of the overall reaction, i.e. whether or not this bond has been broken by the time the reaction mixture attains the transition state for that stage. This may be seen very clearly in, for example, the nitration of benzene (1) in which the attacking species, under normal conditions, has been shown (p. 90) to be the *nitronium ion*, $^{\oplus}NO_2$:

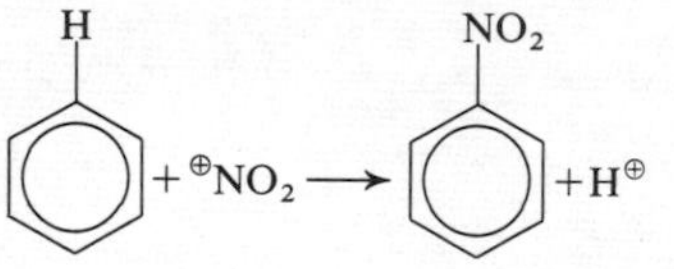

(1)

A C—H bond of the original benzene must clearly be broken at some stage in the overall nitration, but the idealized rate law

$$\text{Rate} = k[C_6H_6][^{\oplus}NO_2]$$

that may be derived from kinetic measurements, offers comparatively little help in deciding at exactly what stage. The above idealized rate

38

law tells us that both the species that figure in it are involved in the transition state for the rate-limiting step of the overall reaction. This certainly rules out a pathway in which the slow, rate-limiting stage of the reaction is fission of the C—H bond before attack by $^{\oplus}NO_2$ takes place, for the rate law would then involve $[C_6H_6]$ only; it does not, however, enable us to decide between at least two alternative interpretations in molecular terms that differ from each other in the stage at which the C—H bond undergoes fission.

Thus the overall nitration could proceed by a one step, concerted pathway (*cf.* $EtO^{\ominus} + MeI$, p. 22) [1],

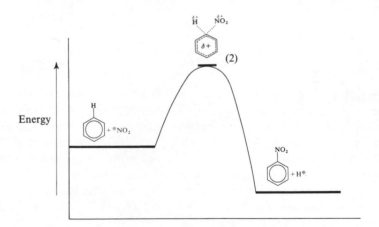

in which the C—H bond is being broken and the C—NO_2 being formed simultaneously; (2) is then a transition state corresponding to the energy maximum in a reaction profile of the form illustrated in Fig. 2.1:

Fig. 2.1

Alternatively, the overall nitration could—because the carbon atom attacked is a quasi-unsaturated one—proceed by a two step pathway in which the new C—NO_2 bond is formed *completely* in a slow, rate-limiting step before the C—H bond begins to undergo fission in a subsequent, fast (non rate-limiting) stage, i.e. [2]

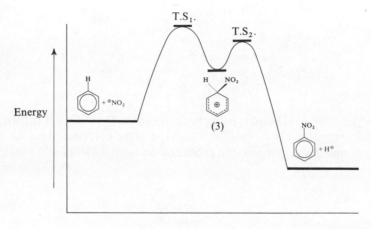

Such a pathway could have a possible energetic advantage in that an energy-yielding step (bond formation) precedes the stage (bond fission) in which energy must be provided. Such a two step pathway may be represented by a reaction profile of the form illustrated in Fig. 2.2,

Fig. 2.2

in which the starting materials are converted, in the slow rate-limiting step, *via* transition state 1 into the *intermediate* (3), which is in turn converted, in a fast, non rate-limiting step, *via* transition state 2 into products. Depending on the depth of the energy 'trough' and the conditions under which nitration is carried out, it is possible with some aromatic species actually to isolate intermediates such as (3) (*cf.* p. 76).

Both these projected pathways are compatible with the experimentally established rate law; they differ from each other in that in [1] the C—H bond is broken in the single—and therefore rate-limiting— stage, while in [2] the C—H bond is not broken until *after* the rate-limiting stage. Simple kinetic measurements alone will not enable us to distinguish between these alternative pathways, but we have established a salient criterion—whether or not the C—H bond is broken in the rate-limiting stage—by which we could at least demonstrate that one of the alternative pathways does not operate in the nitration of benzene, provided we can devise an experimental procedure capable of monitoring the bond fission. This could be done, as we shall see below (p. 42), by comparing the rates of nitration, under comparable conditions, of ordinary benzene (C_6H_6) and of benzene in which the hydrogen atoms have been replaced by the heavier isotope deuterium (C_6D_6).

This is an example of the use of an isotope in a kinetic context to provide information about a reaction pathway, but isotopes can also provide us with otherwise inacessible information about reaction pathways in non-kinetic contexts. Thus in the reaction of simple carboxylate anions (4) with cyanogen bromide (5) at 250°,

$$RCO_2^{\ominus} + BrCN \xrightarrow{250°} RCN + CO_2 + Br^{\ominus}$$
$$\quad (4) \qquad\quad (5) \qquad\quad (6)$$

it seems not unreasonable to suppose that the product cyanide (6) is formed by a simple displacement reaction, and that the carbon dioxide that is also formed derives from the carboxylate group of (4). If, however, the reaction is carried out on an anion (4a) whose carboxylate carbon atom is labelled with the carbon isotope ^{14}C (*cf.* p. 50), it is found that the product carbon dioxide contains no ^{14}C label, as simple displacement would require, all the label being in the cyanide (6a) carbon atom:

$$R^{14}CO_2^{\ominus} + BrCN \xrightarrow{250°} R^{14}CN + CO_2 + Br^{\ominus}$$
$$\quad (4a) \qquad\quad (5) \qquad\qquad (6a)$$

We thus, rather unexpectedly, have to look for a reaction pathway other than simple displacement (*cf.* p. 51). Further non-kinetic uses of ^{14}C, and of the isotopes of other elements, to provide information about reaction pathways will be discussed subsequently (p. 50); but first the kinetic uses of isotopes will be considered in more detail.

KINETIC USES

(a) Primary kinetic isotope effects

In the nitration of benzene referred to above, the C—H bond fission may be monitored by comparing the rates of nitration, under comparable conditions, of C_6H_6 and C_6D_6, because it is found in practice that carbon-deuterium (C—D) bonds are attacked and broken more slowly than are comparable carbon-hydrogen (C—H) bonds. Somewhat inexact calculations based on quantum mechanical considerations suggest a rate ratio, k_{C-H}/k_{C-D}, of ≈ 8 at 25°, and experimentally determined values of this order have been regularly observed (*cf.* p. 43). It thus follows that if the breaking of a C—H bond is involved in the rate-limiting stage of the nitration of benzene (mode [1], above), we should expect to find that C_6D_6 underwent nitration more slowly than C_6H_6; whereas if the breaking of a C—H bond is not so involved (mode [2], above), then we should expect to find no rate difference on nitration of the two. In practice we find experimentally* that the two species undergo nitration at essentially the same rate, C—H bond breaking is thus not involved in the rate-limiting stage of the overall reaction and mode [1]—the concerted pathway—clearly cannot be operating.

The fact that there is no observed rate difference also rules out another possible (i.e. compatible with the experimentally determined rate law) pathway [3] that we have not previously considered: a two-step mode involving the same intermediate (3), but with the first, bond-forming, step fast and the second, bond-breaking, step slow:

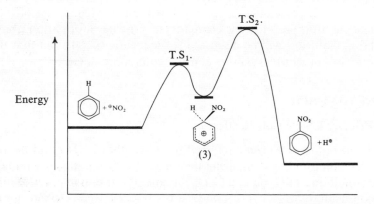

This would, of course, be represented by a reaction profile of the form illustrated in Fig. 2.3:

Fig. 2.3

*The experiments were actually carried out, because of their greater solubility in the nitrating agent and for other practical reasons, on $C_6H_5NO_2$ and $C_6D_5NO_2$, but this does not affect the form or validity of the argument.

We thus see that while three—and quite possibly more—different pathways for nitration were compatible with the initial kinetic evidence, a further kinetic isotopic experiment was able unequivocally to exclude two of them. It is, however, important to emphasize that this does not mean that nitration is thereby *proved* to proceed by pathway [2], we can merely say that this satisfies all the experimental evidence to date and can, therefore, be used as a useful working hypothesis until such time as further evidence may be produced that requires its modification, or even outright rejection.

An essentially analogous problem confronts us in the oxidation of benzaldehyde (7) with neutral permanganate

$$Ph-C{\overset{O}{\Big/}}_{\backslash H} \xrightarrow{MnO_4^{\ominus}} Ph-C{\overset{O}{\Big/}}_{\backslash O^{\ominus}}$$

(7)

where, again, a C—H bond is broken: is it, or is it not, involved in the rate-limiting step of the overall reaction? Synthesis of the mono-deuterated analogue (7a), followed by oxidation under conditions parallel to those employed with benzaldehyde (at 25°)

$$Ph-C{\overset{O}{\Big/}}_{\backslash D} \xrightarrow{MnO_4^{\ominus}} Ph-C{\overset{O}{\Big/}}_{\backslash O^{\ominus}}$$

(7a)

shows that (7a) undergoes oxidation 7·5 times more slowly than benzaldehyde (7), i.e. $k_H/k_D = 7\cdot5$ (at 25°). The breaking of the C—H(D) bond in this case must therefore be involved in the rate-limiting step of the overall reaction, which is said to exhibit a *primary kinetic isotope effect*. Primary here refers to the fact that the isotopic label is itself a constituent of the bond that is being broken, i.e. k_H/k_D refers to the relative rates of fission of the two isotopically different bonds (*cf.* secondary kinetic isotope effects below, p. 49).

The existence of such experimentally detectable primary kinetic isotope effects raises the question of why such an isotopic substitution (i.e. no variation in essential chemical character of the atoms involved in the bond) should be capable of effecting a difference in rates for the parallel processes:

$$Y-H+X \xrightarrow{k_H} Y+H-X$$
$$Y-D+X \xrightarrow{k_D} Y+D-X$$

Each starting material possesses vibrational energy and, so far as the contribution of the Y—H and Y—D bonds, respectively, is concerned, this will differ from one compound to the other. The vibrational energy is related to the frequency of the vibration which, in turn, is related to the masses of the atoms involved in the bond: the greater the masses, the lower the frequency of the vibration and the lower the vibrational energy. The relative vibrational energy levels of the starting materials, Y—H and Y—D, will thus be as represented in Fig. 2.4:

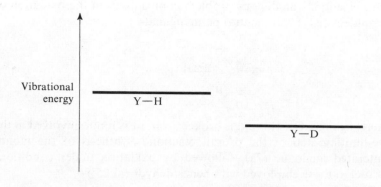

Fig. 2.4

There are, however, no exactly analogous vibrations involving H (or D) in the respective transition states (8 and 9),

$$Y\cdots\cdots H\cdots\cdots X \qquad Y\cdots\cdots D\cdots\cdots X$$
$$(8) \qquad\qquad\qquad (9)$$

the stretching vibration having now been converted, in the transition state, largely into translation along the reaction coordinate, i.e. into increasing the separation of the Y and H (or D) atoms. The highly isotope-sensitive stretching vibration of Y—H (or Y—D) is thus lost on conversion of starting materials into the corresponding transition states, and the effect of isotopic substitution on the activation energy ($E_{Y-H}^{\ddagger}$ or $E_{Y-D}^{\ddagger}$) of the reaction, and hence on its rate, is thus governed very largely by difference of vibrational energy between the two starting materials (Fig. 2.5, p. 45). Because vibrational energy differences between C—H and C—D decrease as the temperature rises we should expect to find that k_{C-H}/k_{C-D} values also decrease with temperature rise: this is indeed observed.

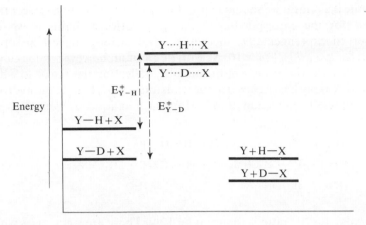

Fig. 2.5

As indicated above (p. 41), somewhat inexact theoretical calculations suggest a k_{C-H}/k_{C-D} value of ≈ 8 (at 25°); values of this order are indeed observed (*cf.* oxidation of PhCDO, p. 43), but so are a number of rather smaller values such as:

$$\frac{k_H}{k_D} = 4\cdot9$$

(10)

$$\text{PhCHCH}_2\overset{\oplus}{\text{NMe}}_3 + {}^{\ominus}\text{OH} \longrightarrow \text{PhCH}{=}\text{CH}_2 + \text{NMe}_3 + \text{H}_2\text{O} \qquad \frac{k_H}{k_D} = 3\cdot0$$

(11)

Such rather smaller values no doubt reflect the fact that the maximum kinetic isotope effect will be observed only when the bond to hydrogen (or deuterium) is *completely* cleaved in the transition state, and its value will decrease the greater the degree of residual bonding to hydrogen (or deuterium) still present at that stage. The actual, experimentally observed, value of a primary isotope effect can thus provide significant information about the nature of the transition state of the reaction in which it occurs.

Kinetic isotope effects may also be observed using the even heavier, radioactive isotope of hydrogen tritium, ${}^3\text{H}$ or T. Because of the greater atomic mass ratio we should expect k_{C-H}/k_{C-T} values to be larger than those for the corresponding k_{C-H}/k_{C-D} ratio, and this is what is observed experimentally; theoretical calculations suggest an upper limit for k_{C-H}/k_{C-T} of ≈ 16. As with deuterium, however, values lower than this are also observed, presumably reflecting the extent to which bond-breaking has occurred in the transition state. Thus in the hydroxyl ion catalysed enolization of the ketone (12) in aqueous solution,

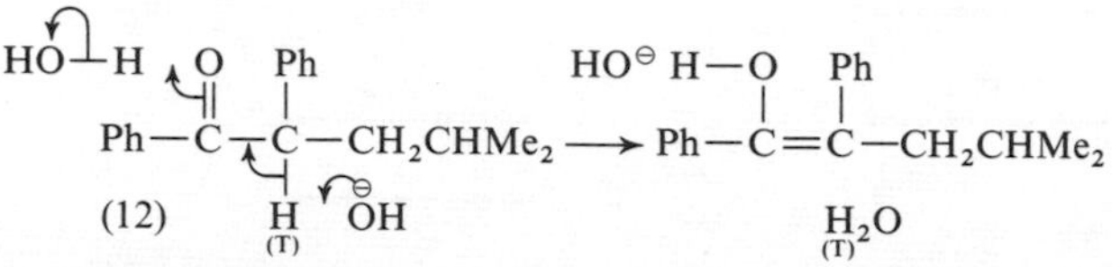

the value for the ratio is found to be 12·9. There are some cases where the larger tritium effects are advantageous, but they require radioactive monitoring and the relevant hydrogen atom (or atoms) is commonly only partly replaced by T in the labelled compound. By contrast deuterium effects do not require radioactive monitoring, and may often be followed directly by ordinary kinetic methods as the relevant hydrogen atom (or atoms) is usually replaced completely by deuterium in the labelled compound. Deuterium kinetic isotope effects involving bonds to atoms other than carbon are also known but are of very much less general utility.

No comment has yet been made about the synthesis of the compounds employed above in which all, or specifically selected, hydrogen atoms have been replaced by deuterium. In a number of cases it is possible to effect the synthesis by direct exchange of deuterium for hydrogen in the already formed compound; this is, for example, the case with benzene (1)

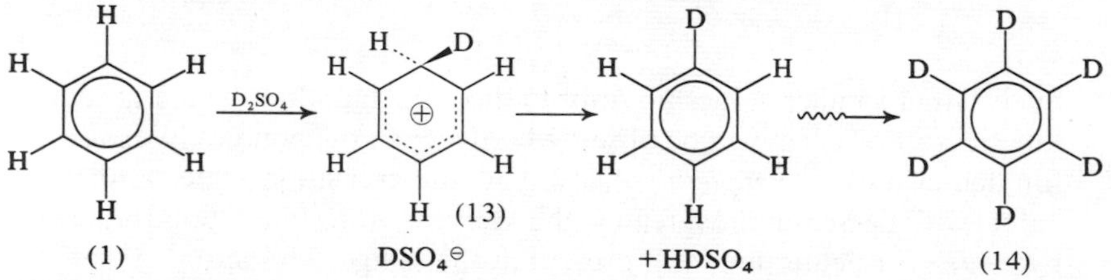

where exposure for a sufficient length of time to a large enough excess of D_2SO_4 results in the successive replacement of all the hydrogen atoms by deuterium, *via* successive σ complexes e.g. (13), to yield the completely deuterated species, benzene-d_6 (14). Somewhat analogously, it is possible to obtain the deutero equivalent of the ketone (12) by base-catalysed exchange of its acidic-hydrogen in a large excess of D_2O:

$$
\begin{array}{ccc}
\underset{\substack{| \\ \text{H} \\ (12)}}{\overset{\substack{\text{O} \quad \text{Ph} \\ \| \quad | \\ \text{Ph}-\text{C}-\text{C}-\text{CH}_2\text{CHMe}_2}}{}} \underset{\text{OD}}{} \rightleftharpoons &
\left[
\begin{array}{c}
\overset{\text{O} \quad \text{Ph}}{\underset{\ominus}{\text{Ph}-\text{C}-\text{C}-\text{CH}_2\text{CHMe}_2}} \\
\updownarrow \\
\underset{+\text{HOD}}{\overset{\ominus\text{O} \quad \text{Ph}}{\text{Ph}-\text{C}=\text{C}-\text{CH}_2\text{CHMe}_2}}
\end{array}
\right] \overset{D_2O}{\rightleftharpoons} &
\underset{\substack{| \\ \text{D} \\ (12a)}}{\overset{\substack{\text{O} \quad \text{Ph} \\ \| \quad | \\ \text{Ph}-\text{C}-\text{C}-\text{CH}_2\text{CHMe}_2}}{}}
\end{array}
$$

Where such a direct exchange is not possible, or could lead to ambiguity about the position occupied by the deuterium atom introduced, then specific synthesis is used often through the agency of deuterium labelled reducing agents. Thus the benzaldehyde singly labelled with deuterium in the aldehyde group (7a) may be obtained from methyl benzoate (15) by use of $LiAlD_4$:

$$
\underset{\substack{| \\ \text{OMe} \\ (15)}}{\text{Ph}-\text{C}=\text{O}} \xrightarrow[\text{2). } D_2O]{\text{1). LiAlD}_4} \underset{\substack{| \\ \text{D}}}{\overset{\substack{\text{D} \\ |}}{\text{Ph}-\text{C}-\text{OD}}} \xrightarrow{\text{oxid.}} \underset{\substack{| \\ \text{D} \\ (7a)}}{\text{Ph}-\text{C}=\text{O}}
$$

The occurrence of primary kinetic isotope effects is not of course confined to bonds containing hydrogen/deuterium, but the very large atomic mass ratio here involved ($1:2$) results in very much larger, and hence more readily detectable, maximum effects than will be observed for other isotopic pairs. An atom that might obviously be expected to interest the organic chemist is carbon, and it has proved possible to detect primary kinetic isotopic effects by use of ^{13}C and ^{14}C. Thus for the reaction of iodomethane (16) with hydroxyl ion (S_N2)

$$
\text{HO}^\ominus + {}^{(14)}_{12}\text{CH}_3-\text{I} \longrightarrow \text{HO}-{}^{(14)}_{12}\text{CH}_3 + \text{I}^\ominus \qquad \frac{k_{^{12}\text{C}-\text{I}}}{k_{^{14}\text{C}-\text{I}}} = 1.09
$$
$$
(16)
$$

the ordinary ^{12}C compound is found to react 1.09 times faster than the ^{14}C labelled compound at $25°$; while in the solvolysis (S_N1) of Me_3C-Cl, the ^{12}C compound is found to react only 1.03 times faster (at $25°$) than the halide in which the tertiary carbon atom is ^{14}C labelled.

These rate differences are sufficiently small that normal titrimetric methods for following the rate of reaction of the labelled compound would hardly be precise enough, even if the labelled compound containing 100 per cent ^{14}C in the relevant carbon atom could be obtained; use is therefore made of the fact that ^{14}C is radioactive. The specific activity of a sample of, for example, iodomethane containing some ^{14}C label (see below, p. 58) is determined, and this figure is subsequently compared with the specific activity of samples of as yet unreacted iodomethane reisolated after the lapse of successive time intervals. It is of course important that the reisolated iodomethane is very pure, i.e. free from contamination with any ^{14}C labelled product methanol, but this may usually be achieved by converting it to a solid derivative followed by exhaustive recrystallization. Comparison of the successive specific activity values, obtained as above, gives a measure of the change in relative isotopic composition of the residual iodomethane, and hence of the relative rates of reaction of the $^{12}CH_3$—I and $^{14}CH_3$—I species.

Experiments with ^{13}C may be handled in an analogous way except that in this case the relative isotopic composition ($^{13}C/^{12}C$) may be determined by mass spectrometry. ^{13}C isotope effects are found to have values about half those of ^{14}C for the same reaction; while this might be expected to make use of the latter preferable, in fact the mass spectrometric measurements of $^{13}C/^{12}C$ relative abundance can be made very much more precisely than the radioactive ones for $^{14}C/^{12}C$ thereby often outweighing the advantage of the latter's absolute magnitude.

Effects have also been observed for isotopes of elements heavier than carbon, though such effects are of course correspondingly smaller. Thus a nitrogen isotope effect has been determined for C—N bond breaking in the elimination reaction of the quaternary compound (11) for which we have already discussed (p. 45) a deuterium effect:

$$PhCH_2CH_2 \overset{(15)}{-} {}^{14}\overset{\oplus}{N}Me_3 + {}^{\ominus}OH \longrightarrow PhCH{=}CH_2 + {}^{14}NMe_3 + H_2O \qquad \frac{k_{14N}}{k_{15N}} = 1{\cdot}01$$
(11)

A sulphur isotope effect has been determined for the similar elimination reaction of the analogous sulphonium compound (17)

$$PhCH_2CH_2 \overset{(34)}{-} {}^{32}\overset{\oplus}{S}Me_2 + {}^{\ominus}OH \longrightarrow PhCH{=}CH_2 + {}^{32}SMe_2 + H_2O \qquad \frac{k_{32S}}{k_{34S}} = 1{\cdot}006$$
(17)

as has a chlorine isotope effect for the solvolysis of benzyl chloride (18):

$$PhCH_2 \overset{(37)}{-} {}^{35}Cl + H_2O \longrightarrow PhCH_2OH + H^{\oplus} {}^{35}_{37}Cl^{\ominus} \qquad \frac{k_{35Cl}}{k_{37Cl}} = 1{\cdot}0076$$
(18)

None of these experimentally determined values is as large as the maximum calculated value for the particular effect, and they presumably reflect the extent of bond fission in the transition state. None of them is remotely as useful in providing information about reaction pathways as are kinetic deuterium, and to a very much lesser extent carbon, effects: the main value of isotopic effects with these other elements is in a non-kinetic context as we shall see subsequently (p. 50).

(b) Secondary kinetic isotope effects

In all the cases we have considered to date the isotope effect involves one or other of the atoms composing the bond that is being broken; the question thus arises whether any kinetic isotope effects have been observed which involve atoms that are not so situated. Among the earliest example investigated was the solvolysis (in 80 per cent ethanol) of the tertiary chloride, 2-chloro-2-methylbutane (19) where replacement of two different groups of hydrogen atoms by deuterium (19a, and 19b) was found to lead to the following reaction rate ratios (isotope effect for the *total* group of deuterium atoms introduced in each case)

$$
\underset{\text{(19a)}}{\overset{\displaystyle \text{Cl} \atop |}{(CH_3)_2CCD_2CH_3}} \qquad \frac{k_H}{k_D} = 1{\cdot}41 \qquad\qquad \underset{\text{(19b)}}{\overset{\displaystyle \text{Cl} \atop |}{(CD_3)_2CCH_2CH_3}} \qquad \frac{k_H}{k_D} = 1{\cdot}78
$$

compared with the undeuterated halide (19). Similarly the rate of solvolysis of 2-chloro-2-methylpropane (20) is found to be slowed by ≈ 10 per cent for each deuterium atom that is introduced. Such effects induced by atoms not involved in the bond undergoing fission are known as *secondary kinetic isotope effects*. They are normally very much smaller than primary effects and, in solvolysis reactions at any rate, tend to occur in S_N1 reactions: in S_N2 reactions they are very much smaller or not detectable at all.

This last observation supplies a clue as to how such secondary effects could operate. Thus in the solvolysis of the halide (20), the slow, rate-limiting step is the formation of the ion pair (21) or, more correctly, the transition state preceding it:

$$
\underset{\text{(20)}}{(CH_3)_3C{-}Cl} \xrightarrow[\text{slow}]{} \text{T.S.} \longrightarrow \underset{\text{(21)}}{(CH_3)_3C^{\oplus}Cl^{\ominus}}
$$

A feature that is believed to stabilize the carbonium ion moiety (*cf.* p. 24) of the ion pair (lower its intrinsic energy level) is the involvement of canonical states such as (21a)

(21a)

this is likely to become progressively less effective as a stabilizing factor with each successive replacement of a hydrogen atom by the heavier deuterium. In so far as (21) can be taken as a model for the transition state that precedes it, this too will be destabilized (will have a higher energy level) as each deuterium atom is introduced; more energy will, therefore, be required to attain the transition state (higher activation energy), and the reaction rate will thus be progressively slowed down. Secondary kinetic isotope effects can, therefore, also supply us with pertinent information about the nature of transition states.

NON-KINETIC USES

It is also possible to make use of isotopic labelling to reveal subtle details of reaction pathways in a non-kinetic context, and here no barrier is imposed—as in the kinetic context—by whether the size of the effect is large enough to make it readily detectable. It is now merely a case of finding a suitable isotope of the element whose behaviour, as a constituent of a bond or group, we wish to study and if necessary— —though this is not always quite so easy—arranging for the synthesis of the desired, selectively labelled compound. Isotopes that have been used in this way include D, ^{3}H(T), ^{13}C, ^{14}C, ^{15}N, ^{18}O, ^{32}P, ^{35}S, ^{37}Cl, ^{131}I and others beside.

One example has already been mentioned (p. 41) in which the use of an isotopic label, in this case ^{14}C, demonstrates that a reaction

$$R^{14}CO_2^{\ominus} + BrCN \xrightarrow{250°} R^{14}CN + CO_2 + Br^{\ominus}$$
(4a)   (5)   (6a)

cannot be the simple displacement of CO_2 from the carboxylate anion (4a) by cyanide, as might otherwise have been tacitly assumed, because the CO_2 evolved is found to contain no ^{14}C label. Any projected pathway for the reaction must, as an essential criterion, require that all the carbon of the product cyanide group be provided by the original carboxylate carbon. It seems likely—though this is not specifically required—that the R—^{14}C bond remains intact throughout the reaction, and a possible pathway acceptable in general chemical terms would thus be:

$$
\begin{array}{ccc}
\text{N} \equiv \text{C} - \text{Br} & \text{N} = \text{C} \overset{\frown}{-} \text{Br} & \text{N} = \text{C} \quad + \text{Br}^{\ominus} \\
\text{R} - {}^{14}\text{C} - \text{O}^{\ominus} \xrightarrow{(5)} \text{R} - {}^{14}\text{C} - \text{O} & \longrightarrow \text{R} - {}^{14}\text{C} \quad \text{O} \\
\underset{(4a)}{\overset{\|}{\text{O}}} & \underset{}{\overset{|}{\text{O}^{\ominus}}} & \overset{\|}{\text{O}} \quad (22)
\end{array}
$$

$$
\begin{array}{ccc}
& & \overset{|||}{} \\
\text{N} \quad \text{C} = \text{O} & \text{N}{-}\text{C} = \text{O} & \text{N} = \text{C} = \text{O} \\
\underset{(6a)}{\overset{|||}{R} - {}^{14}\text{C}} \quad \overset{\|}{\text{O}} & \longleftarrow \text{R} - {}^{14}\text{C} - \text{O} & \longleftarrow \text{R} - {}^{14}\text{C} = \text{O} \quad (22)
\end{array}
$$

This is an essentially conjectural pathway, but it receives considerable support from the fact that it has proved possible, in some cases, to isolate intermediates such as (22), i.e. RCONCO.

A number of different elements will now be discussed in turn to illustrate their utility as labels.

(a) Oxygen-18

Almost certainly the earliest non-kinetic use of an isotopic label to investigate a reaction pathway involved water enriched in ^{18}O, i.e. containing some molecules of $H_2{}^{18}O$, in the hydrolysis of amyl acetate (23):

$$
\underset{(23)}{\text{Me} - \text{C} \overset{\text{O}}{\underset{\text{OC}_5\text{H}_{11}}{}} } + H_2O \rightleftharpoons \text{Me} - \text{C} \overset{\text{O}}{\underset{\text{OH}}{}} + \text{HOC}_5\text{H}_{11}
$$

The hydrolysis of the ester could clearly involve either acyl-oxygen fission [1] or alkyl-oxygen fission [2]

$$
\underset{(23)}{\text{Me} - \text{C} \overset{\text{O}}{\underset{[1] \quad \text{O} \overset{[2]}{\mid} \text{C}_5\text{H}_{11}}{}}} + H_2O \rightleftharpoons \text{Me} - \text{C} \overset{\text{O}}{\underset{\text{OH}}{}} + \text{HOC}_5\text{H}_{11}
$$

but no kinetic measurement will enable us to distinguish between them. If, however, water enriched in ^{18}O is employed then the ^{18}O label will be found in the product acetic acid if acyl-oxygen fission, [1], occurs and in the product amyl alcohol if alkyl-oxygen fission, [2], operates:

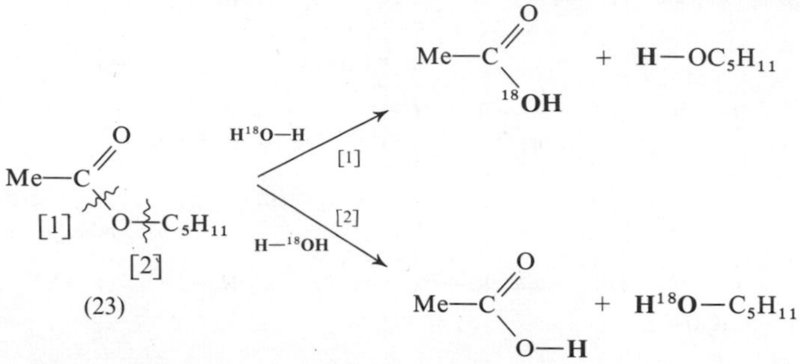

Clearly, if both modes of fission are in operation simultaneously then ^{18}O will be found in both acid and alcohol. After careful chemical separation of acid and alcohol it was found in fact that no ^{18}O label at all was present in the amyl alcohol and the hydrolysis, under these conditions, must thus proceed entirely *via* [1], acyl-oxygen fission.

The results in this case are quite unequivocal, but there are two important general requirements that must be satisfied in experiments of this kind if such is to be the case. The first is that the product acid and alcohol, once formed, should not undergo any subsequent ^{18}O exchange with the residual ^{18}O enriched water; this can indeed be shown to be the case here. The second is that the separation of acid and alcohol should be meticulous so that any possibility of residual contamination of one by traces of the other can be ruled out; here, at least, this too presents no problem. There remains, however, the question of how enrichment in ^{18}O may actually be detected in one or other of the separated products. At the time the above experiment was carried out (1934) this was done by combustion of the product followed by very precise determination of the density of the water that resulted: today it is much more conveniently done directly by mass spectrometry.

Water enriched in ^{18}O has also been used to investigate the hydration equilibrium of carbonyl compounds:

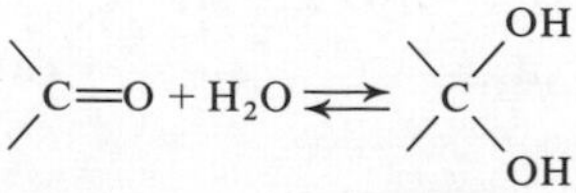

In the case of formaldehyde and acetaldehyde hydration may be detected spectroscopically, and the equilibrium position shown to correspond to ≈ 100 per cent hydrate (24) formation with the former and to 58 per cent with the latter. By contrast, no equilibrium concentration at all of hydrate (24) can be detected under analogous conditions with acetone (25), and the question therefore arises as to whether such a hydration equilibrium is operative in this case. Dissolving acetone (25) in ^{18}O enriched water, in the presence of a trace of acid or base as catalyst, is found to result in rapid incorporation of ^{18}O into the carbonyl group:

$$Me_2\overset{\delta+}{C}=\overset{\delta-}{O} + H_2^{18}O: \rightleftharpoons Me_2C\overset{OH}{\underset{\overset{18}{O}H}{\big\langle}} \rightleftharpoons Me_2C={}^{18}O + H_2O:$$

$$\qquad (25) \qquad\qquad\qquad (24a)$$

The hydration equilibrium is thus shown to operate with acetone (25) as with the simple aldehydes, but is not directly detectable in the same way because the ambient hydrate (24a) concentration is so very small.

Another area in which information may be gained from ^{18}O isotopic labelling is, hardly surprisingly, that of oxidation and, in particular, over the question of whether the oxygen atom (or atoms) in an oxidation product has been provided by an oxygen-containing oxidizing agent, an oxygen-containing solvent, e.g. water, or both. Thus in the oxidation of benzaldehyde (26) with aqueous permanganate, under acid or neutral, conditions it may be shown that one of the atoms of oxygen in the product benzoic acid (27) is provided by ^{18}O enriched permanganate ion:

$$Ph-C\overset{O}{\underset{H}{\big\langle}} + Mn^{18}O_4^{\ominus} \xrightarrow{H_2O} Ph-C\overset{O}{\underset{\overset{18}{O}H}{\big\langle}}$$

$$\qquad (26) \qquad\qquad\qquad (27)$$

This may appear to be stressing the obvious—or at least the expected—but that things are not always quite so cut and dried is suggested by the fact that if the above experiment is repeated under alkaline conditions then only very little ^{18}O label is now introduced into the product benzoic acid (27): the added oxygen must therefore be provided, under these conditions, very largely by the solvent water! It should perhaps be emphasized that when ^{18}O is incorporated into the benzoic acid it is, of course, not possible to determine whether it is incorporated as 'carbonyl' or 'hydroxyl' oxygen because of the operation of the dynamic equilibrium:

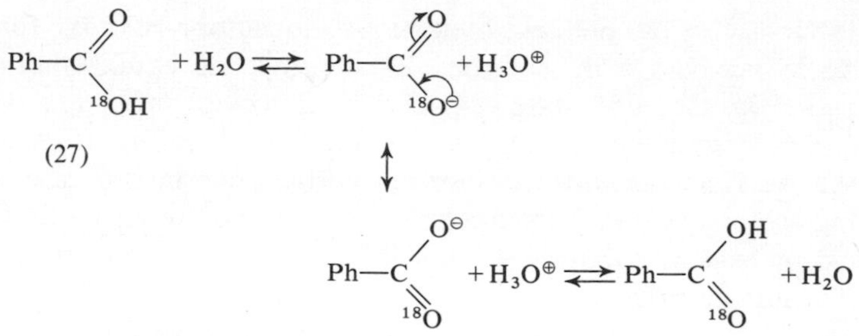

$$Ph-C\overset{O}{\underset{^{18}OH}{\diagdown}} + H_2O \rightleftharpoons Ph-C\overset{\diagup O}{\underset{^{18}O^{\ominus}}{\diagdown}} + H_3O^{\oplus}$$

(27)

$$Ph-C\overset{O^{\ominus}}{\underset{^{18}O}{\diagdown}} + H_3O^{\oplus} \rightleftharpoons Ph-C\overset{OH}{\underset{^{18}O}{\diagup}} + H_2O$$

(27a)

Another example of the use of $Mn^{18}O_4^{\ominus}$ to provide information about an oxidation reaction and, by inference, about an intermediate involved is discussed subsequently (p. 151); in a further example, ^{18}O labelling is used to investigate the stereochemistry of an intermediate (p. 229).

(b) Deuterium

The widespread non-kinetic use of deuterium as an isotopic label stems in no small measure from the ready availability of D_2O. An example of its use is in the Cannizzaro reaction with benzaldehyde (28)

$$2PhCHO + {}^{\ominus}OH \xrightarrow{H_2O} PhCO_2{}^{\ominus} + PhCH_2OH$$
(28)   (29)

where the question arises as to whether the second atom of hydrogen in the CH_2 of the product benzyl alcohol (29) is provided by another molecule of the initial benzaldehyde (28), by a molecule of the solvent water, or—much less likely—by a hydroxyl ion. Carrying out the reaction in D_2O (which also means that the base will now be $^{\ominus}OD$) is found to result in the incorporation of *no* deuterium into the methylene group of the benzyl alcohol (29), though the hydroxyl group is found to contain D. The second atom of hydrogen in the methylene group must thus be provided by another molecule of benzaldehyde (28), and transferred in such a way that it does not, at any stage, become free— as $H^{\oplus}$, $H\cdot$ or $H^{\ominus}$—for it would then be able to exchange with the very large excess of D_2O, and it would be the resultant $D^{\oplus}$, $D\cdot$ or $D^{\ominus}$ that would be incorporated into the methylene group of the benzyl alcohol (29). This information, coupled with the rate law

$$Rate = k[PhCHO]^2[{}^{\ominus}OH]$$

deduced from experimental kinetic data, suggests a possible reaction pathway

$$Ph-\overset{\overset{\ominus OD}{|}}{\underset{\underset{O}{\|}}{C}}-H \underset{fast}{\rightleftharpoons} Ph-\overset{\overset{OD}{|}}{\underset{\underset{O^{\ominus}}{|}}{C}}-H \quad \overset{\overset{O}{\|}}{\underset{\underset{H}{|}}{C}}-Ph \underset{slow}{\longrightarrow} Ph-\overset{\overset{OD}{|}}{\underset{\underset{O}{\|}}{C}} \;+\; H-\overset{\overset{\ominus O}{|}}{\underset{\underset{H}{|}}{C}}-Ph$$

(28)　　　(30)　　　(28)

$$\downarrow fast$$

$$Ph-\overset{\overset{O^{\ominus}}{|}}{\underset{\underset{O}{\|}}{C}} \;+\; H-\overset{\overset{DO}{|}}{\underset{\underset{H}{|}}{C}}-Ph$$

(29a)

in which reversible addition of $^{\ominus}OD$ (or $^{\ominus}OH$) to a molecule of benzaldehyde (28) is followed by transfer of H, with its electron pair (i.e. effectively as $H^{\ominus}$), *directly* from this adduct (30) to a second molecule of benzaldehyde in the slow, rate-limiting step of the reaction. The transition state thus involves two molecules of benzaldehyde and one hydroxyl ion as the rate law demands, and the reaction is completed by rapid proton (deuteron) transfer—almost certainly *via* the solvent— to convert the initial acid + alkoxide ion into the more stable pair, carboxylate ion + alcohol.

Further support for the suggested reaction pathway is provided by running the complementary reaction on benzaldehyde (28a) in which the aldehyde hydrogen is deuterium-labelled (*cf.* p. 47), but this time in water:

$$2PhCDO + {}^{\ominus}OH \xrightarrow{H_2O} PhCO_2{}^{\ominus} + PhCD_2OH$$
　　　(28a)　　　　　　　　　(29a)

As the suggested pathway would require, all the deuterium label is found in the methylene group of the product alcohol (29a). In both the above cases, the occurrence or non-occurrence of deuterium in the methylene group can readily be detected by i.r. spectroscopy (C—H stretching in CH_2 at 2850 cm^{-1}, C—D stretching in CD_2 at 2120 cm^{-1}), or n.m.r. spectroscopy (C—H in CH_2 at τ 5·4, C—D in CD_2 shows no signal).

(c) Nitrogen-15

This isotope of nitrogen is a non-radioactive, naturally occurring one that may be enriched in a number of simple nitrogenous compounds by distillation, diffusion etc; one such compound that is particularly readily available in highly enriched form (95 per cent+) is ammonia, i.e. $^{15}NH_3$. Use has been made of this for labelling experiments in connection with amino acid and peptide chemistry and also, of course, with nitrogen-containing heterocyclic compounds. Thus the quaternary isothiazolium salt (31) is found on treatment with ammonia to undergo dealkylation to yield the corresponding isothiazole (32):

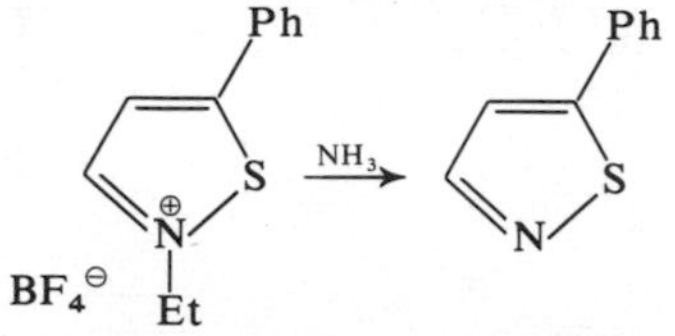

(31)   (32)

This was, not unnaturally, assumed to be a simple transfer of the ethyl group from the isothiazolium nitrogen atom to the attacking ammonia molecule—a process familiar from numerous other examples—until the experiment was repeated with $^{15}NH_3$ (actually ammonia 97·8 per cent enriched in ^{15}N). It was then found, most surprisingly, that the product isothiazole (32a) not only contains ^{15}N label but contains it in essentially undiluted form:

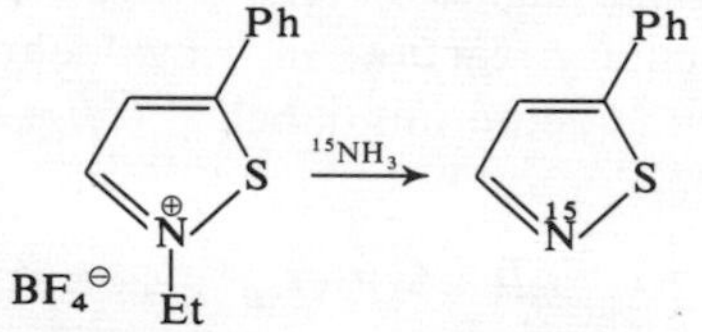

(31)   (32a)

The reaction thus cannot proceed by simple, direct dealkylation but must presumably involve the addition of NH_3, opening of the ring, and its subsequent reclosure with selective elimination of the original ring nitrogen atom—perhaps by the following pathway:

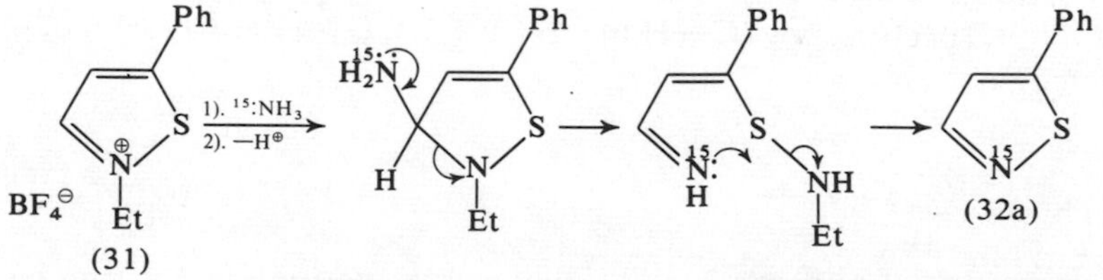

(31)

The extent of incorporation of ^{15}N in the isothiazole (32a) may be readily determined by mass spectrometry on the highly purified product, and comparison of the relative intensities of the m/e peaks at 161 and 162, corresponding to the molecular ions from (32, ^{14}N) and (32a, ^{15}N), respectively.

(d) Sulphur-35

In contrast to the nitrogen isotope above, ^{35}S is radioactive with a half-life of 87·1 days, thus making it very convenient for use in experiments of the usual duration of a few hours. An interesting example of its employment is in an investigation of the interconversion (*cf.* p. 36) of the isomeric naphthalene 1- and 2-sulphonic acids (33, and 34, respectively):

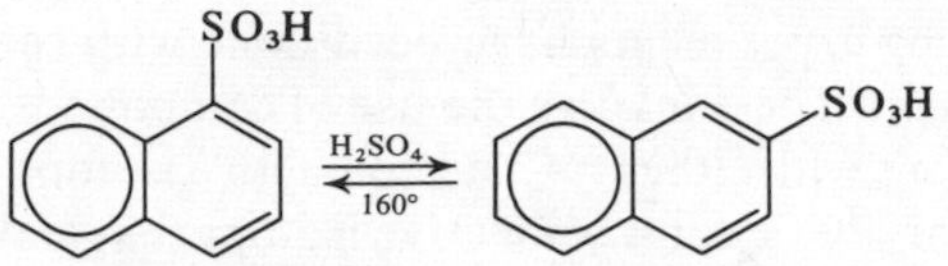

(33)   (34)

Either pure sulphonic acid is converted, on heating in concentrated H_2SO_4 at 160°, into the same (equilibrium) mixture containing 80 per cent of (34) and 20 per cent of (33); the question then arises as to whether this takes place by direct migration of the SO_3H group from one position to the other (presumably without its becoming free in the process), or by *desulphonation* to naphthalene (35) followed by *resulphonation* to yield the other isomer:

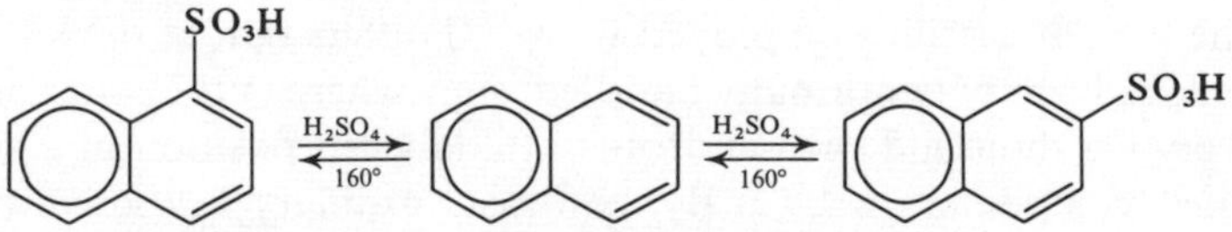

(33)   (35)   (34)

It should be possible to establish which of these alternative pathways is operative by carrying out the equilibration of either pure sulphonic acid at 160° in $H_2{}^{35}SO_4$. If equilibration proceeds by desulphonation/resulphonation then ^{35}S label should become incorporated in the product sulphonic acids from the $H_2{}^{35}SO_4$ during resulphonation; while if it proceeds by direct migration of the SO_3H group from one position to the other then no incorporation of ^{35}S should take place. It is chastening but instructive to find, on experiment, that ^{35}S label *is* incorporated into the product sulphonic acids, but that this occurs *more slowly* than the rate at which their interconversion takes place: in other words, the theoretically definitive isotopic labelling experiment does not in fact provide us with the expected clear-cut answer! There could be a number of different explanations of what is actually observed experimentally: that both the suggested reaction pathways operate simultaneously, or that after desulphonation to naphthalene (35) some resulphonation takes place by the departing H_2SO_4, because of its proximity, rather than by the more numerous, surrounding molecules of $H_2{}^{35}SO_4$; and other explanations are possible—the detailed answer is not yet known.

(e) Carbon-13 and carbon-14

So many bond-making and bond-breaking processes in organic chemistry involve carbon that it comes as no surprise to find that more isotopic labelling experiments have been done with carbon than with any other element. These involve the use of compounds labelled in one or more positions with either ^{14}C, a radioactive isotope with a half-life of 5730 years, or ^{13}C, a non-radioactive isotope that may be monitored by mass spectrometry as has already been mentioned (p. 48). The β-radiation emitted by ^{14}C is such that while it is readily detected and counted, it does not constitute a major hazard, precautions need to be taken in handling compounds labelled with it but this can normally be achieved without elaborate laboratory adaptation etc. ^{14}C also has the advantage of so long a half-life that no corrections need to be made for natural decay even in experiments of extended duration, and labelled compounds that are otherwise chemically stable can be stored over considerable periods of time before use.

The level of labelling employed with ^{14}C differs but, at the levels of specific radioactivity normally handled, somewhere in the order of one in every ten thousand carbon atoms at a labelled position in a singly labelled compound is ^{14}C. By contrast, ordinary carbon already contains approximately one atom of ^{13}C in every hundred in the natural state and this level has to be raised if it is to be used as a label; the degree of enrichment may, in some cases, rise to one in every two carbon atoms at a labelled position. Both ^{14}C and ^{13}C—but particularly the former—have been used in a much wider variety of investigations of reaction pathways than any other isotopes, and a number of examples of these will now be considered.

(i) Inter- *v.* intra-molecularity of rearrangements:

In molecular rearrangements it is often important to know whether, in the course of the structural change, the two fragments resulting from bond-breaking ever actually become free, i.e. have an existence independent of each other, before rejoining (bond-formation) to form the rearranged product, or whether they don't (*cf.* the naphthalene sulphonic acids example considered above): in the former case the rearrangement is said to be *intermolecular*, in the latter *intramolecular*. One such example that has been the subject of a good deal of study is the acid-catalysed rearrangement of hydrazobenzene (36) to benzidine (37), generally referred to as the *benzidine rearrangement;* this is undergone by a wide variety of N,N′-diarylhydrazines:

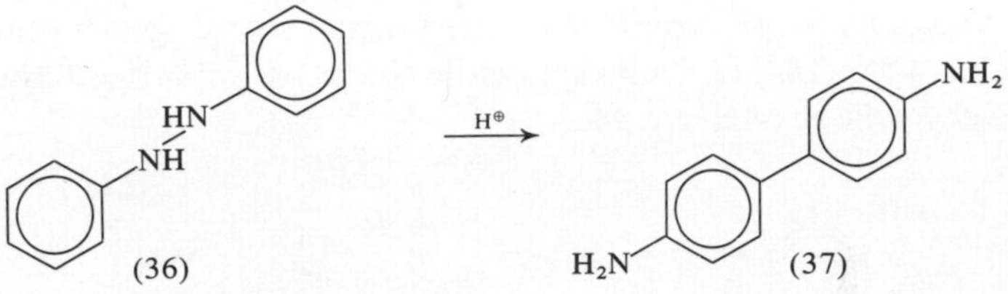

The usual test for inter- or intra-molecularity is a 'crossover' experiment (*cf.* p. 108) in which two different, though closely similar, compounds (38 and 39), that separately rearrange at comparable rates, are allowed to rearrange together, at comparable concentrations, in the same solution. This is then examined for the presence, in addition to the normal rearrangement products (40 and 41, respectively), of any of the mixed (crossover) product (42):

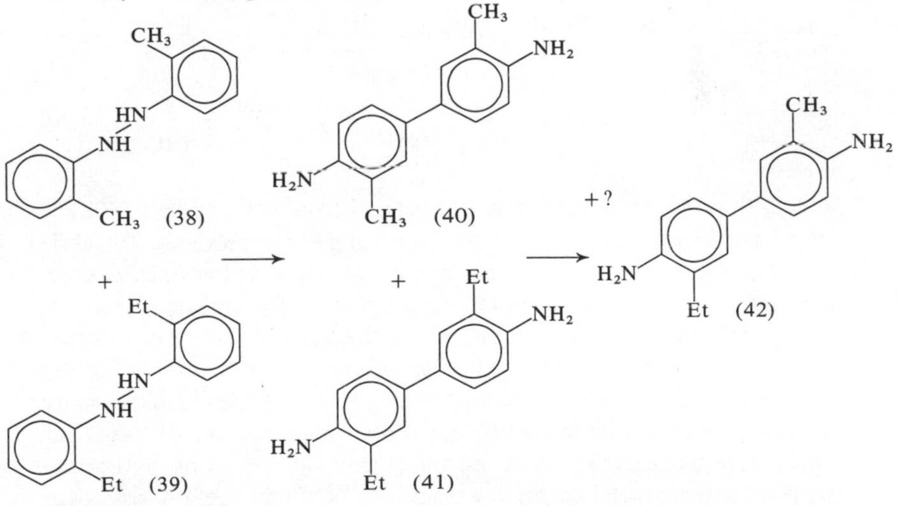

If any crossover product (42) is found to be present then the rearrangement must, in part at least, be intermolecular, i.e. the fragments produced from breaking the N—N bonds in (38) and (39) must enjoy some degree of independent existence if they are to be able to combine one with another rather than merely to recombine with each other. In practice no crossover product can be detected in the above case, but this is not necessarily conclusive as it could be due to failure to detect a product that might only be present in small relative proportion anyway, and in the presence of large quantities of two other compounds of very closely similar properties.

This particular dilemma may be overcome by rearrangement, in the same solution, of (38) and the unsymmetrical ^{14}C labelled N,N'-diarylhydrazine (43), from which the normal rearrangement products will be (40 and 44, respectively), while the potential mixed (crossover) products will be (40a and 44a):

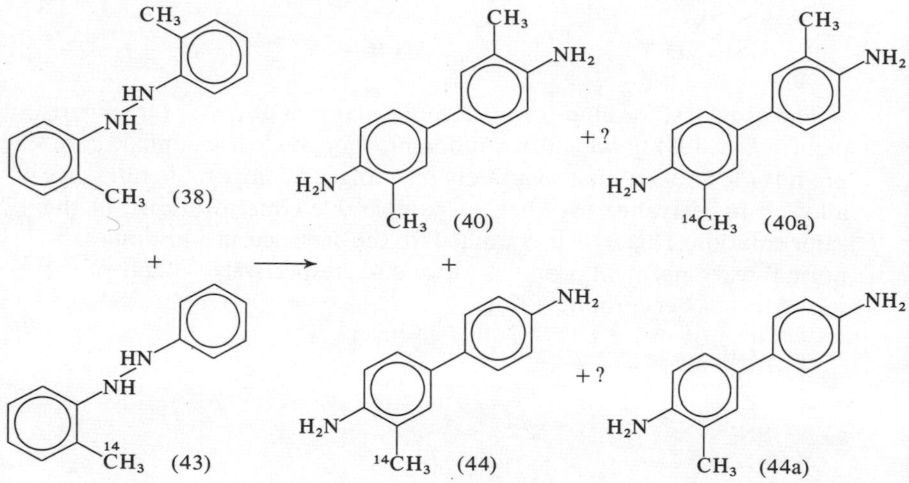

CH₃ ... HN–NH ... CH₃ (38)

+

HN–NH ... ¹⁴CH₃ (43)

⟶

CH₃ ... NH₂ ... H₂N ... CH₃ (40)

+

H₂N ... NH₂ ... ¹⁴CH₃ (44)

+ ?

CH₃ ... NH₂ ... H₂N ... ¹⁴CH₃ (40a)

+ ?

H₂N ... NH₂ ... CH₃ (44a)

It will thus be seen that one potential crossover product (40a) is chemically identical with the normal rearrangement product (40), while the other (44a) is chemically identical with the second normal rearrangement product (44). No attempted separation of a minor constituent is thus necessary, all that is required is the simple (column chromatography) separation of (40) and (44), followed by measurement of the specific activity of each. It is found in practice that the specific activity of (44) is identical with that of (43), i.e. no dilution of the ^{14}C label has taken place on rearrangement, and also that (40) contains, within the limits of experimental error, no trace of ^{14}C label, i.e. no crossover products have been formed and the rearrangement is unequivocally intramolecular. Some other aspects of this famous, not to say notorious, rearrangement, will be discussed subsequently (p. 216).

(ii) Determination of relative migratory ability:

In the benzilic acid rearrangement of a benzil derivative (45) in which one of the benzene rings carries a substituent

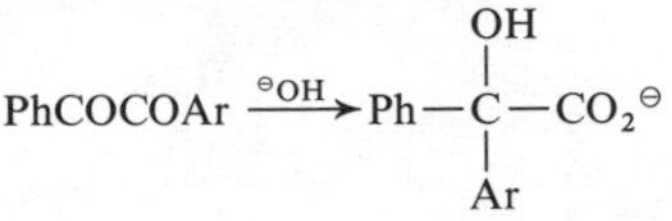

$$PhCOCOAr \xrightarrow{\ominus OH} Ph-\overset{\overset{\displaystyle OH}{|}}{\underset{\underset{\displaystyle Ar}{|}}{C}}-CO_2{}^{\ominus}$$

$$\quad (45) \qquad\qquad\qquad (46)$$

the resultant benzilic acid anion (46) could result from the migration of either of the aryl residues:

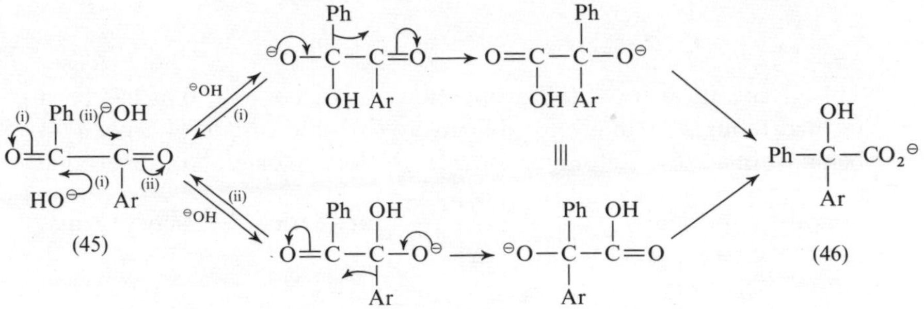

The question thus arises whether the same group always migrates or, as seems more likely, if each group migrates (in separate molecules of course) then what are the relative migratory aptitudes of Ph and Ar, i.e. how much of the total product (46) is produced by migration of Ar as compared with the amount produced by migration of Ph?

This question can be answered by the synthesis and rearrangement of the ^{14}C labelled diketone (45a), for the resultant benzilic acid anions (46a and 46b) will now differ in the position in which they carry their label, depending on which group, Ar or Ph, migrated during their formation:

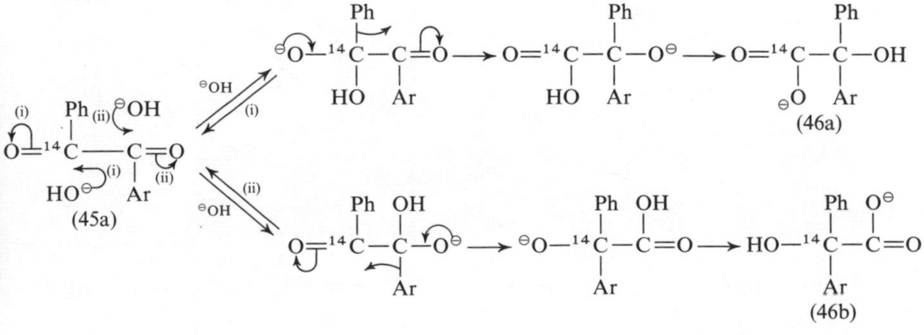

The product benzilate anion (46a + 46b) is then degraded, in the cold, to PhCOAr (47) which is isolated, purified, and its specific activity compared with that of the initial diketone (45a):

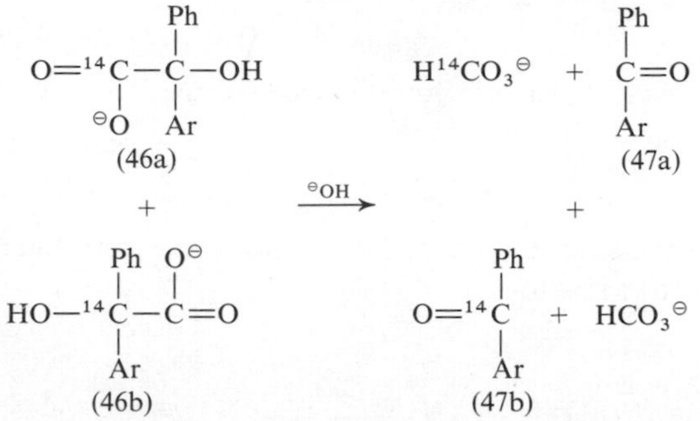

$$O={}^{14}C-\underset{\underset{Ar}{|}}{\overset{\overset{Ph}{|}}{C}}-OH \qquad H^{14}CO_3^{\ominus} \;+\; \underset{\underset{Ar}{|}}{\overset{\overset{Ph}{|}}{C}}=O$$

$$\underset{\ominus O}{}$$

(46a)   (47a)

$$+ \qquad \xrightarrow{\;\ominus OH\;} \qquad +$$

$$HO-\underset{\underset{Ar}{|}}{\overset{\overset{Ph}{|}}{{}^{14}C}}-\overset{\overset{O^{\ominus}}{|}}{C}=O \qquad\qquad O={}^{14}\underset{\underset{Ar}{|}}{\overset{\overset{Ph}{|}}{C}} \;+\; HCO_3^{\ominus}$$

(46b)   (47b)

This gives a measure of the proportion of (47, i.e. 47b) that has been formed though Ar migration (pathway (ii)), the rest of (47, i.e. 47a) being formed by Ph migration. It is thus possible to determine migratory aptitudes $\left(\dfrac{\%\,\text{migration of Ar}}{\%\,\text{migration of Ph}}\right)$ for a series of Ar substituents:

Ar	M.A.
$p\text{-MeC}_6\text{H}_4$	0·63
$o\text{-MeC}_6\text{H}_4$	0·03
$p\text{-ClC}_6\text{H}_4$	2·05
$m\text{-ClC}_6\text{H}_4$	4·10
$o\text{-ClC}_6\text{H}_4$	0·46
$p\text{-MeOC}_6\text{H}_4$	0·46

(iii) Information about possible intermediates:

This kind of information can often be obtained as a result of seeking—through carbon labelling experiments—to follow the fate of a particular, and salient, carbon atom throughout the course of a reaction. A good example is seen in the Claisen rearrangement of phenolic allyl ethers (e.g. 48):

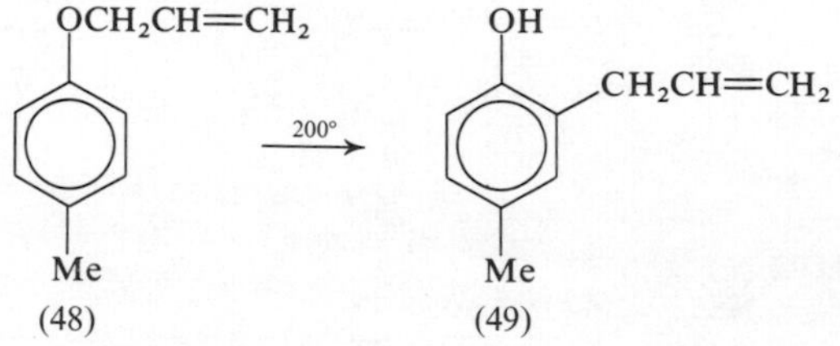

(48)   (49)

This rearrangement has been shown to be intramolecular by attempted, but wholly unsuccessful, crossover experiments of the type described above (p. 59); the question remains, however, as to whether it is the carbon atom (of the allyl group) originally bonded to oxygen in (48) that becomes bonded to the carbon atom of the benzene nucleus in (49) or not? This point could naturally be decided by labelling that carbon atom itself, but it can equally well be decided by labelling another carbon atom in the molecule whose spatial relationship to the first is known. This possibility of alternative sites for the label is important because the major determinants in such a tracer study are very often the relative ease of unequivocal synthesis of the starting material, i.e. the unambiguous introduction of a labelled carbon atom into a known position, and the relative ease of detection of the label's new position subsequently, i.e. after rearrangement. On this basis it was therefore decided to synthesize the compound (48a) in which the ^{14}C label is located in the terminal position—*non*-oxygen end—of the allyl group (the % yields in each step are specified):

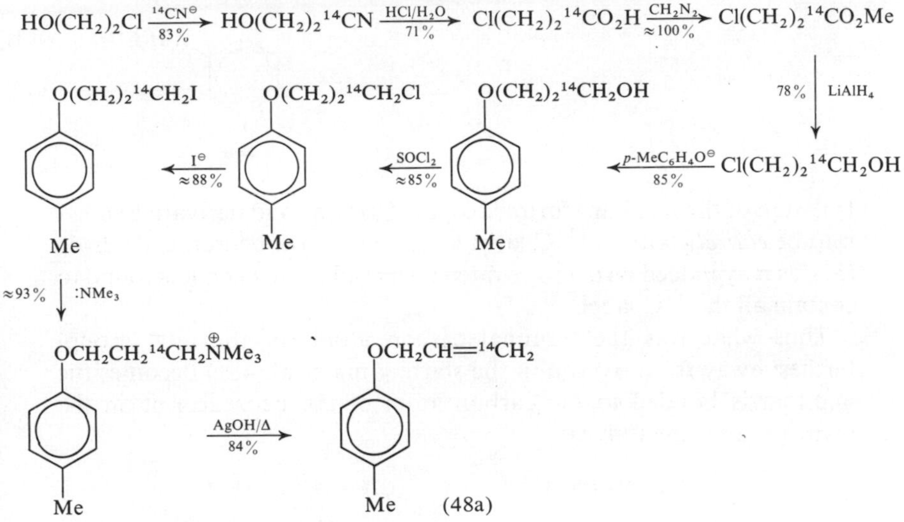

$$HO(CH_2)_2Cl \xrightarrow[83\%]{^{14}CN^\ominus} HO(CH_2)_2{}^{14}CN \xrightarrow[71\%]{HCl/H_2O} Cl(CH_2)_2{}^{14}CO_2H \xrightarrow[\approx100\%]{CH_2N_2} Cl(CH_2)_2{}^{14}CO_2Me$$

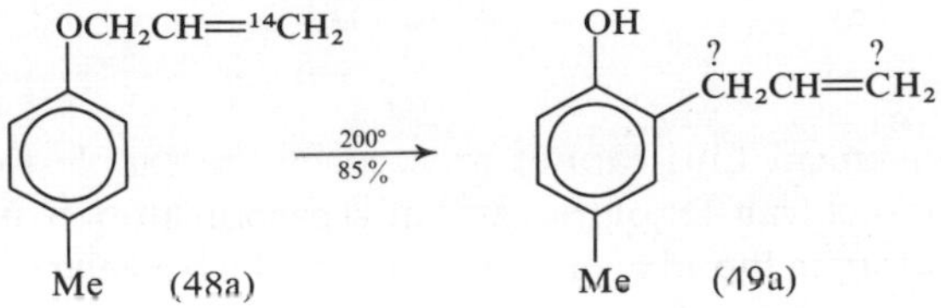

Rearrangement of (48a) is then found to yield the dialkyl phenol (49a) in 85 per cent yield but we have still to determine where the ^{14}C label is located in the rearranged product (49a). It is likely to be at C_1 or C_3 of the allyl group (those shown with a question mark in 49a above) depending on which end of the latter has formed the new bond to carbon in the benzene nucleus; a choice may be made between them by carrying out the following selective oxidative fission of the molecule (the initial methylation of the phenolic hydroxyl group is carried out to prevent the unselective oxidation, in the benzene nucleus of the free phenol, that would otherwise result):

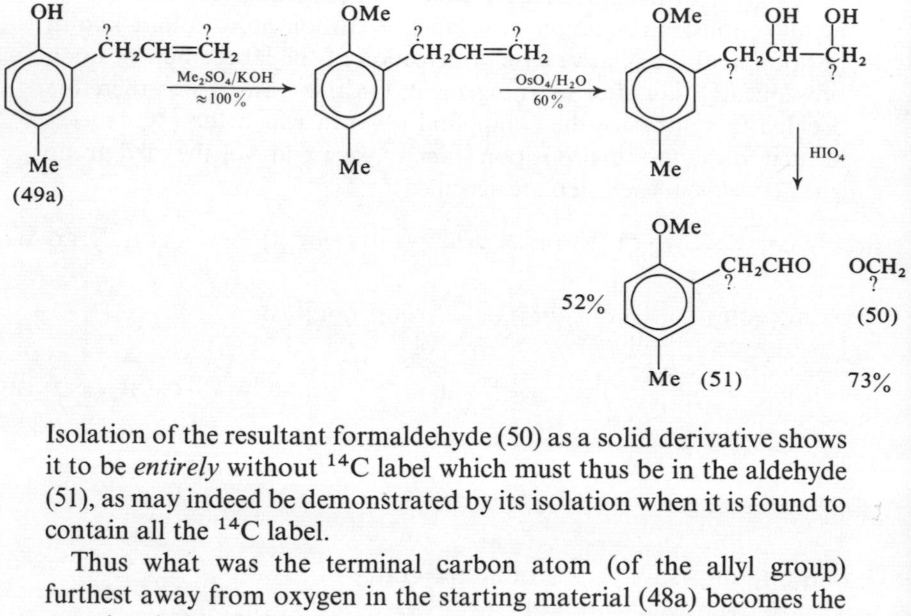

Isolation of the resultant formaldehyde (50) as a solid derivative shows it to be *entirely* without ^{14}C label which must thus be in the aldehyde (51), as may indeed be demonstrated by its isolation when it is found to contain all the ^{14}C label.

Thus what was the terminal carbon atom (of the allyl group) furthest away from oxygen in the starting material (48a) becomes the one that is bonded to the carbon atom of the benzene ring in the rearranged product (49b):

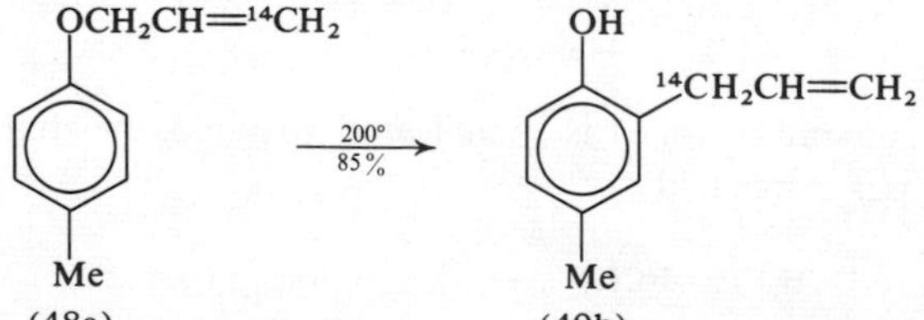

The rearrangement thus cannot proceed by the simplest conceivable pathway—direct transfer of the same allyl carbon atom from oxygen to the carbon atom in the adjacent *o*-position of the benzene nucleus—and the pathway most in accord with present data is something of the form:

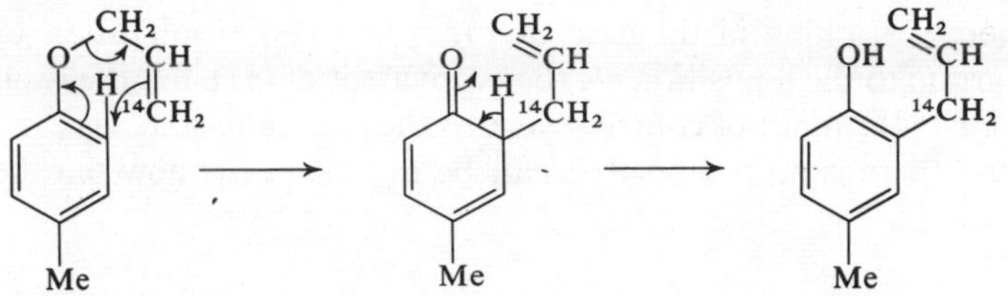

This seems a not unreasonable suggestion in that the specific conformation required in (48a) above is one that the side chain could readily take up in terms of bond angles, bond lengths etc; there is also an energetic advantage in that the energy needed for breaking the C—O bond in (48a) could, in part, be provided by the simultaneous formation of the C—C bond in (49b) *via* a sterically acceptable, six-membered, cyclic transition state (51):

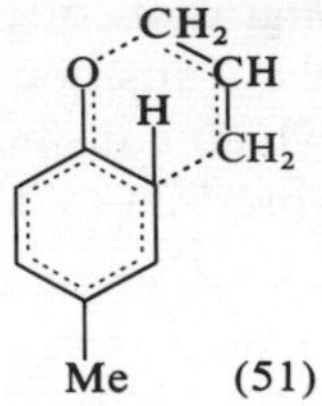

The fact that no 'scrambling' of the ^{14}C label takes place at all constitutes extremely cogent additional evidence for the intramolecularity of the rearrangement, for if the allyl group were ever free one would have expected its subsequent bonding to the benzene ring to proceed through either terminal carbon with consequent scrambling of the ^{14}C label.

If both *o*-positions in an O-allylphenol are blocked (e.g. 52), then, on heating, the allyl group is now found to migrate to the *p*-position (53). If, however, the original O-allylphenol (52) carries a ^{14}C label in the terminal (*non*-oxygen end) carbon of its allyl group (52a, *cf.* 48a), then on rearrangement the label is, in contrast to (49b) above, still found to be *wholly* in the terminal carbon of the allyl group in the rearranged product (53a):

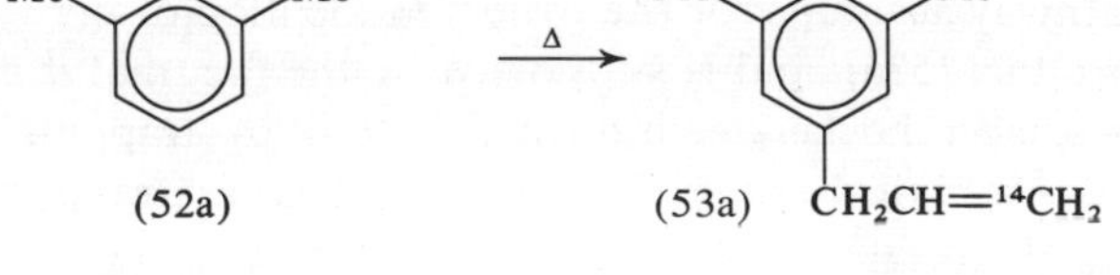

The reaction pathway in this case cannot therefore proceed via the p-equivalent of (51).

There is nothing in the case of (52a) to prevent migration of the allyl group to the o-position via the equivalent of (51), but the resultant dienone (54) cannot of course revert to the aromatic state, as can (50), because there is no proton that can be lost from the now tetrahedral o-position:

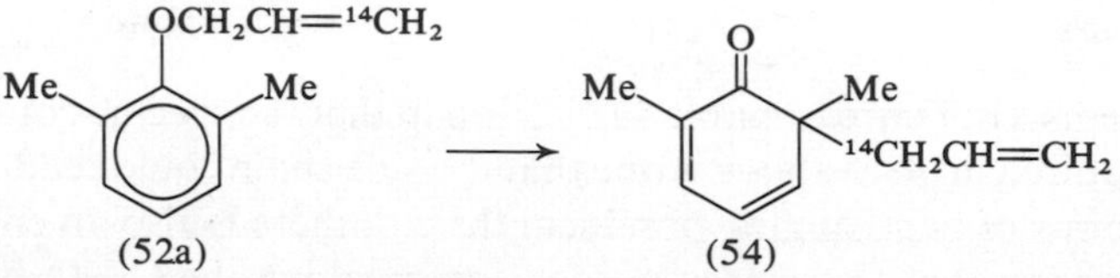

$$OCH_2CH={}^{14}CH_2$$

Me ⟶ Me (52a)

Me ⟶ Me, O, ${}^{14}CH_2CH=CH_2$ (54)

The allyl group is now, however, within 'spanning' distance of the p-position which does have a proton to lose. The available data could then be accounted for by migration of the allyl group in (54) to the p-position, via a second cyclic transition state (55), to yield a second dienone intermediate (56), which can finally lose a proton (cf. 50) to form the fully aromatic end product (53a):

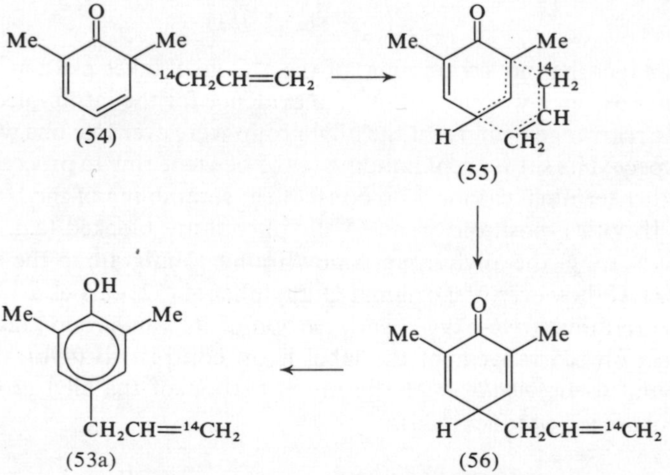

Me, O, Me, ${}^{14}CH_2CH=CH_2$ (54) ⟶ Me, O, Me, ${}^{14}CH_2$, CH, CH_2, H (55)

(55) ↓ Me, O, Me, H, $CH_2CH={}^{14}CH_2$ (56)

Me, OH, Me, $CH_2CH={}^{14}CH_2$ (53a) ⟵ (56)

Such a pathway accounts for the observed location of the ${}^{14}C$ label in the end product (53a), and is supported by the fact that it has proved possible to detect the dienone intermediate (54) by 'trapping' it (p. 104) during the course of the overall (52a) ⟶ (53a) rearrangement.

(iv) Detection of a symmetrical intermediate:

In contrast to the above example where no 'scrambling' at all of the label takes place, cases are known where the scrambling is found to be total, i.e. a label that was confined to one carbon atom in the starting material is found to be spread uniformly throughout all the carbon atoms of the product. This is, for example, the case when cyclo-heptatriene (57), ^{14}C labelled in the saturated carbon (photochemically induced synthesis as shown), is oxidized with CrO_3 to yield benzoic acid (58):

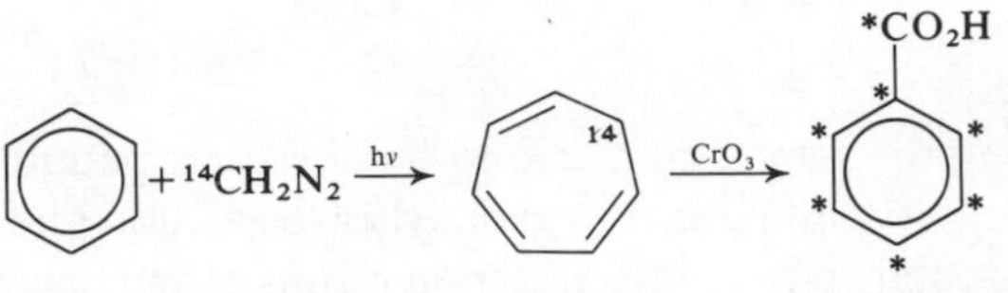

(57)　　　　(58)

The pattern of scrambling of the ^{14}C label in the product (58) is charted by its degradation in a Schmidt reaction (with hydrazoic acid, HN_3) to carbon dioxide and aniline (59):

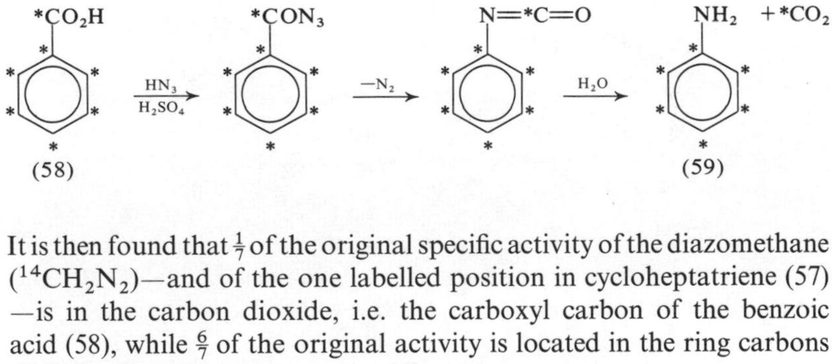

(58)　　　　　　　　　　　　　　　　　　(59)

It is then found that $\frac{1}{7}$ of the original specific activity of the diazomethane ($^{14}CH_2N_2$)—and of the one labelled position in cycloheptatriene (57) —is in the carbon dioxide, i.e. the carboxyl carbon of the benzoic acid (58), while $\frac{6}{7}$ of the original activity is located in the ring carbons of the aniline (59).

It is important to be clear exactly what scrambling means in this context; thus (58) has been written, for convenience, with an asterisk (*) against each carbon, but this does not mean that *any* single molecule of (58) contains 7 ^{14}C atoms. Most of the total number of molecules of (58) will of course contain no ^{14}C at all, and those that do contain ^{14}C will not contain more than one atom of it, i.e. we are talking about a mixture of non- and mono-labelled compounds, the latter having the single ^{14}C label in different situations:

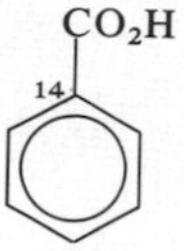

CO_2H

(58a)

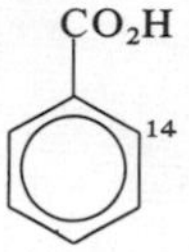

CO_2H

(58b)

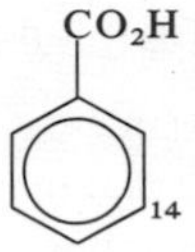

CO_2H

(58c)

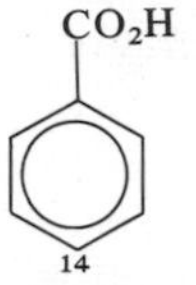

CO_2H

(58d)

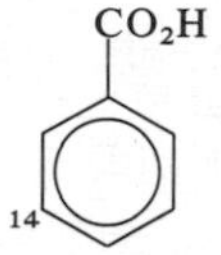

CO_2H

(58e)

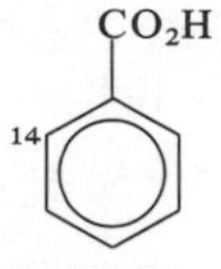

CO_2H

(58f)

$^{14}CO_2H$

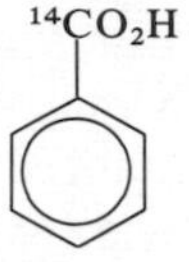

(58g)

Thus considering the sum total of mono-labelled compounds, $\frac{1}{7}$ of the original total activity will be in the carboxyl carbon atom and the other $\frac{6}{7}$ of the original total activity in the carbon atoms of the benzene rings, i.e. scrambling is here a statistical spread of the original activity among all the available carbon atoms in the product.

This can only have come about by all these 7 carbon atoms having been equivalent at some stage in the oxidation of (57) ⟶ (58), i.e. the reaction must have proceeded *via* a symmetrical intermediate. The suggested candidate is the tropylium cation (60) which is well known in other contexts:

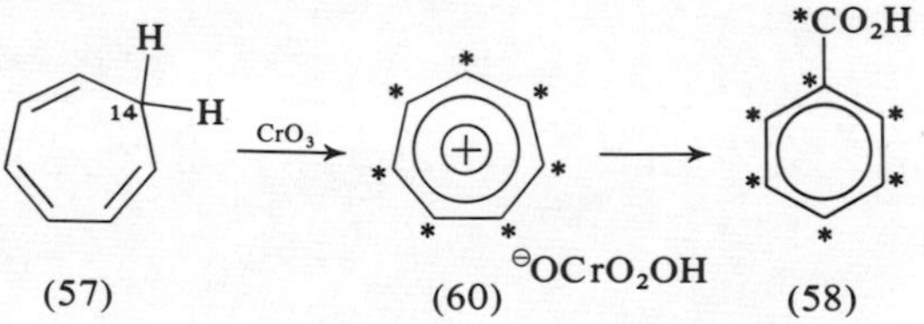

$\xrightarrow{CrO_3}$

(57)

$^{\ominus}OCrO_2OH$

(60)

$^{*}CO_2H$

(58)

Somewhat less elaborate cases of scrambling are also known, in which the fact that as few as two carbon atoms have been equivalent at some stage in an overall reaction (*cf.* p. 209) throw useful light on the structure of potential intermediates.

(v) Biogenetic and biodegradative uses:

A further and most important use of carbon isotopic labelling has been in investigating the synthetic and degradative abilities of living organisms, and of enzyme systems. It can certainly be said that the extent to which the mechanism of photosynthesis by green plants has been unravelled stems almost entirely from the use of $^{14}CO_2$ to trace the succession of molecular species through which it undergoes transformation into carbohydrates. Carbon isotopic labelling has also been used to trace processes in the opposite direction, i.e. the metabolism of foreign molecules, such as drugs or potential chemotherapeutic remedies, into successively smaller molecular species to ensure that potentially toxic metabolic products are not produced *en route* to the species that are ultimately excreted.

The largest detailed use of carbon isotopic labelling has, however, been in the field of biogenesis—in seeking to trace the pathways by which plants elaborate large and complex molecules such as fatty acids, steroids, alkaloids, antibiotics etc. from simple starting materials. A particularly fruitful example has been the use of labelled acetate (61), for acetate has long been known to act as an intermediate in a wide variety of biochemical processes. Acetate is first coupled with coenzyme-A and then, in this active form (62), is capable of undergoing a variety of C—C bond forming operations with itself and with other simple metabolic products. By feeding ^{14}C labelled acetate (61) to a suitable plant, mould etc. and then, by normal chemical degradative reactions (*cf.* p. 64), locating the position of ^{14}C labelled carbon atoms in molecules isolated subsequently from it, general biosynthetic pathways may be charted. An example in point is the antibiotic griseofulvin (63) which is elaborated by one of the strain of the fungus *Penicillium griseofulvum*,

$$CH_3{}^{14}CO_2H + CoA{-}SH \longrightarrow CH_3{}^{14}COS{-}CoA \rightsquigarrow$$

(61)   (62)

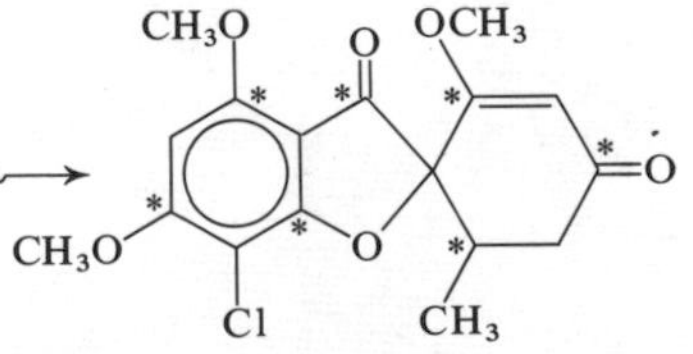

(63)

the product is found to contain no less than 7 labelled carbon sites—located as shown above—and it is possible, from their relative positions, to work out a potential biosynthetic scheme which may then be tested further.

In the course of this chapter intermediates have been mentioned on a number of occasions and it is now proposed to go on and study them in detail.

FURTHER READING

ARNSTEIN, H. R. V. and BENTLEY, R. 'Isotopic Tracer Technique', *Quart. Rev.,* 1950, **4**, 172.

BURWELL, R. L. 'Deuterium as a Tracer in Reactions of Hydrocarbons on Metallic Catalysts', *Acc. Chem. Res.,* 1969, **2**, 289.

COLLINS, C. J. 'Isotopes and Organic Reaction Mechanisms', *Adv. Phys. Org. Chem.* (Academic Press), 1964, **2**, 1.

COLLINS, C. J. and BOWMAN, N. S. (Eds.) *Isotope Effects in Chemical Reactions* (Van Nostrand, 1970).

GOLD, V. 'Applications of Isotope Effects', *Chem. in Britain,* 1970, **6**, 292.

HALEVI, E. A. 'Secondary Isotope Effects', *Progr. Phys. Org. Chem.,* 1963, **1**, 109.

MELANDER, L. *Isotope Effects on Reaction Rates* (Ronald Press, 1960).

RAAEN, V. F., ROPP, G. A. and RAAEN, H. P. *Carbon-14* (McGraw Hill, 1968).

SAMUEL, D. and SILVER, B. L. 'Oxygen Exchange Reactions of Organic Compounds', *Adv. Phys. Org. Chem.* (Academic Press), 1965, **3**, 123.

SAUNDERS, W. H. 'Kinetic Isotope Effects', *Surv. Progr. Chem.* (Academic Press), 1966, **3**, 109.

SIMON, H. and PALM, D. 'Isotope Effects in Organic Chemistry and Biochemistry', *Angew. Chem. Int. Ed.* 1966, **5**, 920.

THOMAS, S. L. and TURNER, H. S. 'The Synthesis of Isotopically Labelled Organic Compounds', *Quart. Rev.,* 1953, **7**, 407.

WESTHEIMER, F. H. 'The Magnitude of the Primary Kinetic Isotope Effect for Compounds of Hydrogen and Deuterium', *Chem. Rev.,* 1961, **61**, 265.

ZOLLINGER, H. 'Hydrogen Isotope Effects in Aromatic Substitution Reactions', *Adv. Phys. Org. Chem.* (Academic Press), 1964, **2**, 163.

3
The study of reactive intermediates

A necessary corollary of the fact that most organic reactions are found to be not elementary (one-step) but complex (multi-step) is that they must proceed through one or more intermediates (Fig. 3.1):

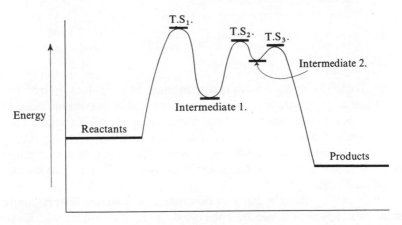

Fig. 3.1

Useful, possibly salient, information about the overall reaction pathway may be obtained by detecting and identifying such intermediates.

This may be done simply by investigation of the reaction mixture to see what is present (but see comment below), or by using kinetic and other evidence to postulate intermediates of specified structure, and then seeking to establish that such are indeed present and involved.

The stability—and hence chance of isolation—of an intermediate is dictated by the depths of the minima in the above energy profile. If the intermediate is at a low energy level compared to the transition state maxima on each side of it, then there is a reasonable chance that it may be stable enough to be isolated. If, as often happens, this is not the case and the intermediate corresponds to only a shallow dip in the energy profile it is likely to enjoy only a transient existence. Isolation is now most unlikely but it may still be possible to detect such an intermediate by physical means, e.g. spectroscopy, or by introducing into the reaction mixture a species with which it might be expected to react particularly readily—*trapping*, as this is called.

It should, however, be emphasized that because a species can be isolated from a reaction mixture this does not necessarily make it an intermediate: it could merely be an alternative end product of the reaction. Thus in the hydrolysis of 2-chloro-2-methylpropane (1), 2-methylpropene (2) may be isolated from the reaction mixture in addition to 2-methylpropan-2-ol (3):

This alkene (2) seems an unlikely intermediate—it does indeed accumulate as the reaction proceeds and is present in maximum amount when all of (1) has reacted—but its formation does provide support for the ion-pair (4) as the real intermediate. Thus attack of $^{\ominus}OH$ (or H_2O:) on the carbonium ion in (4) at (*a*) carbon ($^{\ominus}OH$ acting as nucleophile) could lead to (3), while attack at (*b*) hydrogen ($^{\ominus}OH$ acting as a base) could lead to (2).

An essential criterion that any postulated, or isolated, intermediate must satisfy is that it can be converted, under the conditions of the reaction, into the normal products at a rate at least as fast as that of the overall reaction under the same conditions. This still does not constitute absolute proof that a particular species is functioning as an intermediate, but it does provide pretty strong supporting evidence.

MAJOR TYPES OF INTERMEDIATE

A number of species have been suggested, and indeed detected, sufficiently often as intermediates in organic reactions to establish them as being of general validity, so that their occurrence no longer occasions undue surprise. Among these are carbonium ions, $R_3C^\oplus$, carbanions, $R_3C^\ominus$, both usually occurring as constituents of ion pairs, radicals, $R_3C\cdot$, carbenes, $R_2\ddot{C}$ (i.e. with no net charge, but highly electron-deficient as the carbon atom has only six electrons in its outer shell), benzynes,

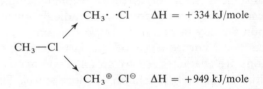

and, rather less commonly, nitrenes, $R\ddot{N}$. In addition there are, of course, a number of other species of less general validity, characteristic of only a small group of reactions or even of a single one.

In the light of experience, certain general guide-lines have emerged that enable us to make at least some guess as to the general type of intermediate that may, or may not, be involved in a particular reaction. Thus reactions proceeding in the gas phase or in solvents of low polarity do not usually involve polar intermediates such as ion pairs. The major reason for this is that it is usually more energy-demanding, in the gas phase, or in solvents of low polarity, to break a bond *heterolytically* to form an ion pair then *homolytically* to form radicals. Thus for chloromethane the relevant ΔH values in the gas phase are

$$CH_3\cdot\ \cdot Cl \qquad \Delta H = +334\ \text{kJ/mole}$$
$$CH_3—Cl$$
$$CH_3^\oplus\ Cl^\ominus \qquad \Delta H = +949\ \text{kJ/mole}$$

while ΔH for the heterolytic fission of chloromethane in aqueous solution has been calculated as only $+263$ kJ/mole; the very considerable difference is due to the large amounts of energy evolved by the solvation of the ion-pair in the highly polar solvent, water. The difference is even more marked for $(CH_3)_3C—Cl$ whose reactions do indeed proceed, in polar solvents, through an ion-pair as we have already seen (p. 23). This assistance in bond-breaking from polar solvents cannot operate in the gas phase or in, for example, hydrocarbon solvents so under these conditions the probability of reactions proceeding through radical intermediates greatly increases.

Reactions involving radical intermediates are attended by a number of other features that distinguish them from those involving polar species. Thus they generally require *initiation* either by addition of substances that are known to produce radicals particularly readily e.g. peroxides, such as benzoyl peroxide

$$\underset{\substack{\text{Ph}-\text{C}-\text{O}\{-\text{O}-\text{C}-\text{Ph}}}{\overset{\text{O}\quad\quad\text{O}}{\parallel\quad\quad\parallel}} \longrightarrow \underset{\substack{\text{Ph}-\text{C}-\text{O}\cdot \;\cdot\text{O}-\text{C}-\text{Ph}}}{\overset{\text{O}\quad\quad\text{O}}{\parallel\quad\quad\parallel}}$$

or by heating

$$\text{PbEt}_4 \overset{\Delta}{\longrightarrow} \text{Pb} + 4\text{Et}\cdot$$

or photochemically:

$$\text{Br}-\text{Br} \overset{h\nu}{\longrightarrow} \text{Br}\cdot \;\cdot\text{Br}$$

Once initiated, such reactions are often very rapid because a self-sustaining chain reaction is set up, the attack of an initiator radical on a molecule of starting material generating another radical that continues the process, for example in the bromination of methane:

$$\begin{array}{c}\text{Br}-\text{Br} \\ \downarrow h\nu \\ \text{CH}_3-\text{H} + \text{Br}\cdot \longrightarrow \text{CH}_3\cdot + \text{H}-\text{Br} \\ \uparrow \qquad\qquad \downarrow \text{Br}-\text{Br} \\ \text{Br}\cdot \;+\; \text{CH}_3-\text{Br}\end{array}$$

As a result of this they are highly susceptible to the addition of *inhibitors*—substances such as phenols, diphenylamine etc., that themselves react particularly readily with radicals—which will stop a chain reaction already in progress, or prevent one from starting in the first place. As a consequence of the latter circumstance many radical reactions are characterized by an *induction period*—a lapse of time after initiation before the reaction proper starts. This is due to added or generated radicals being selectively scavenged by any inhibitor substances already present in the reaction mixture until the latter are saturated, only then will any radicals become available to propagate the reaction proper.

ISOLATION OF INTERMEDIATES

Undoubtedly the most convincing evidence for the participation of a species as an intermediate in a reaction is its isolation, and the subsequent demonstration that it may be converted into end-products at a rate at least as fast as the overall reaction under the same conditions; this requires that it correspond to a reasonably deep energy 'hollow' in Fig. 3.1 (p. 71).

A reaction that is documented pretty well in this respect is the Hofmann reaction of amides

$$R-\overset{\overset{\textstyle O}{\|}}{C}-NH_2 \xrightarrow[\ominus OH]{Br_2} RNH_2$$

Under suitable conditions it is possible, with some amides, to isolate RCONHBr, $RCONBr^{\ominus}\ Na^{\oplus}$, and $RN = C = O$, and to show that each is indeed a real intermediate. This, quite unusually, establishes several fixed points that any projected reaction pathway must traverse. With so much data available from intermediates it is possible to suggest a reaction pathway as follows:

$$\underset{H_2N}{\overset{R}{\diagdown}}C{=}O \xrightarrow[\ominus OH]{Br_2} \underset{HNBr}{\overset{R}{\diagdown}}C{=}O \xrightarrow{\ominus OH} \underset{\underset{\displaystyle Br}{N^{\ominus}}}{\overset{R}{\diagup}}C{=}O \quad (5)$$

$$\downarrow \text{slow}$$

$$RNHCO_2H \xleftarrow{H_2O} R{-}N{=}C{=}O \longleftarrow R{-}\overset{\ominus}{N}{-}\overset{\oplus}{C}{=}O$$

$$\qquad\qquad\qquad\qquad\quad (7) \qquad\qquad\qquad (6)$$

$$\downarrow \ominus OH$$

$$RNH_2 + HCO_3^{\ominus}$$
$$(8)$$

Kinetics tell us that the slow, rate-limiting stage of the reaction is the step involving loss of bromide ion from (5). There is good, though not perhaps conclusive, evidence that this loss and the migration of the R group occurs simultaneously, rather than *via* a two-step process involving a carbonylnitrene, $RC-\ddot{N}$. This stems from such information as the fact that none of the stable products that might be expected to arise from attack of other species present in the reaction mixture on $RC\ddot{O}\ddot{N}$, e.g. $H_2O: \longrightarrow RCONHOH$, has ever been detected. That the migrating R group goes directly from $C \longrightarrow N$ without ever becoming free is demonstrated by the fact that in an optically active compound (5a), the R group suffers no change of configuration on migration (*cf*. p. 165):

$$\underset{\underset{\displaystyle Br}{N^{\ominus}}}{\overset{\overset{\textstyle R}{\diagdown}}{\underset{R''}{\diagup}}}C{=}O \longrightarrow \underset{R''}{\overset{\overset{\textstyle R}{\diagdown}}{\diagup}}C{-}\overset{\ominus}{N}{-}\overset{\oplus}{C}{=}O \rightsquigarrow \underset{R''}{\overset{\overset{\textstyle R}{\diagdown}}{\diagup}}C{-}NH_2$$

$$\qquad (5a) \qquad\qquad\qquad\qquad\qquad\qquad\qquad\qquad (8a)$$

The hydration of isocyanates such as (6) to carbamic acids (7) is well-known, as is the very rapid loss of carbon dioxide by the latter in alkaline solution to yield the product amine (8).

(a) Aromatic electrophilic substitution

The Hofmann reaction is quite unusual in that it is possible to isolate several, successive intermediates; the more common situation is that one is pretty lucky to be able to isolate even one! Even to do this, it may be necessary to use a reagent slightly different from the normal one, or to carry out the reaction under rather different conditions; particularly by reducing the temperature, where this is possible, for an intermediate is likely to be relatively more stable at lower temperatures.

A good example of these constraints is provided, in aromatic electrophilic substitution, by the nitration of trifluoromethylbenzene (9) to its *m*-nitro derivative with nitrosyl fluoride, NO_2F, and boron trifluoride (a somewhat unusual nitrating agent). If the reaction is carried out at $< -50°$, however, a solid separates which, on raising the temperature above $-50°$, decomposes without melting to yield the normal end product in quantitative yield. This solid, which can be shown to have structure (10), is clearly a likely intermediate:

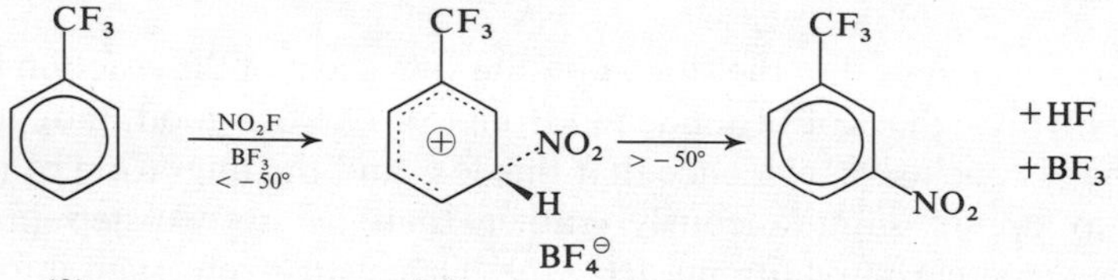

Somewhat less exotic, though still at reduced temperature, is the Friedel Crafts alkylation of 1,3,5-trimethylbenzene (11) with ethyl fluoride as alkylating agent and boron trifluoride as the Lewis acid catalyst which, at $-80°$, yields the intermediate (12) as an orange solid with m.p. $-15°$:

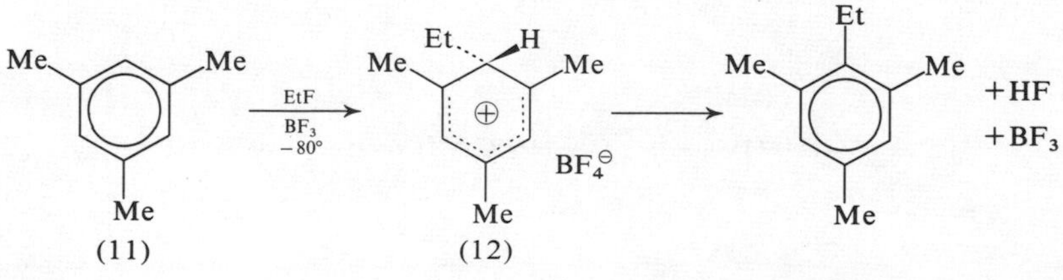

Here again, raising the temperature results in conversion of the inter-mediate (12) into the normal alkylated end product in very high yield. Further, the rate of conversion of intermediate into end product is at least as fast as the rate of the overall reaction under comparable conditions—an essential criterion for an intermediate to fulfil.

This does not of course prove that similar, though less stable, inter-mediates are involved in all the more familiar examples of aromatic electrophilic substitution, e.g. the nitration of benzene with nitrating mixture; but given that the formation of such an intermediate—generally called a σ complex—is compatible with the kinetic evidence (as it is), and that it is supported by isotopic experiments (*cf.* p. 41), it does at least make such a supposition entirely reasonable. In the above examples there is some reason to believe that the particular, and peculiar, stability of the σ complex intermediates stems in part from the very stable $BF_4^{\ominus}$ anion in the ion-pair. That $BF_4^{\ominus}$ is not, however, an essential feature is demonstrated by the formation of a very stable σ complex in the exhaustive methylation of benzene, or methylated benzenes, with chloromethane and aluminium chloride as Lewis acid catalyst:

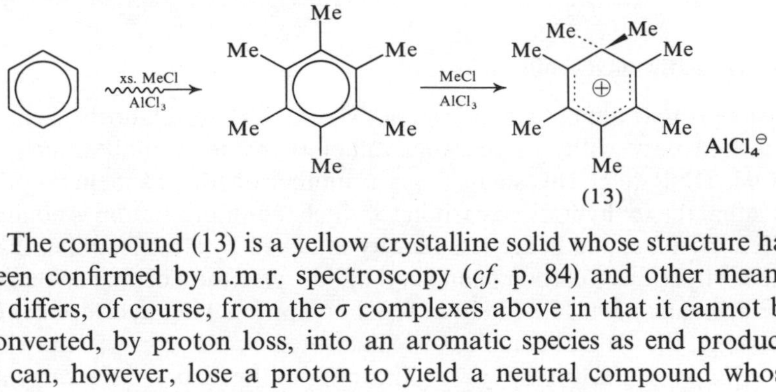

(13)

The compound (13) is a yellow crystalline solid whose structure has been confirmed by n.m.r. spectroscopy (*cf.* p. 84) and other means. It differs, of course, from the σ complexes above in that it cannot be converted, by proton loss, into an aromatic species as end product. It can, however, lose a proton to yield a neutral compound whose structure has been shown by mass, and n.m.r., spectroscopy to be (14):

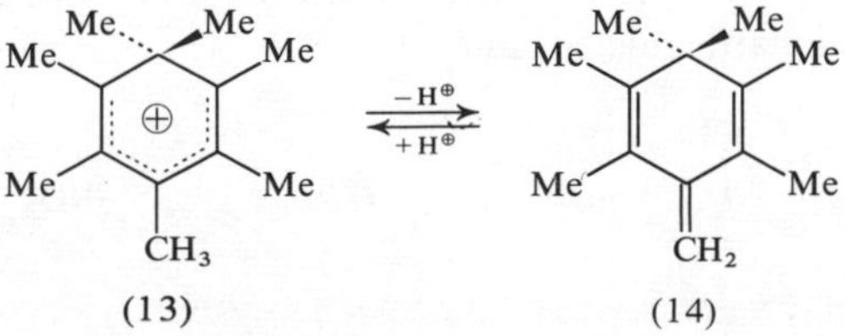

(13)   (14)

The stability of (13) is attested by the fact that (14) is so basic (i.e. ready to take up a proton) that it may be extracted from pentane into aqueous hydrochloric acid (provided the latter is >4M): very surprising behaviour for an unsaturated hydrocarbon.

Even more surprising σ complexes may be generated by working in the 'super-acid' systems of Olah, thus with SbF_5 in liquid SO_2 the following bicyclic species (15) and (16) may be detected by n.m.r. spectroscopy:

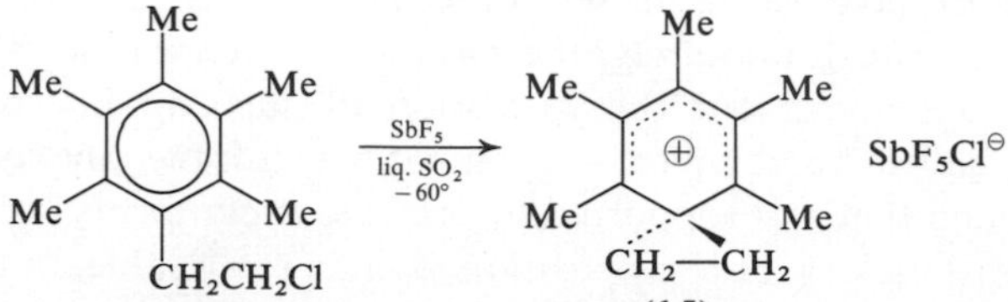

(15)

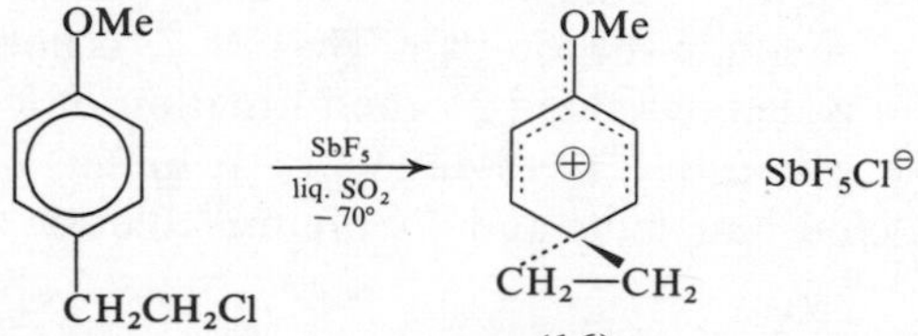

(16)

(b) Aromatic nucleophilic substitution

Unlike simple aliphatic halides, e.g. $MeCH_2CH_2Cl$, chlorobenzene is attacked only with considerable difficulty by nucleophiles such as $^{\ominus}OH$, $H\ddot{N}R_2$ etc. This stems from a number of reasons including the fact that the sp^2 hybridized carbon to which the chlorine atom is bonded is less positive, and hence less readily attacked by a nucleophile, than the sp^3 hybridized carbon of an alkyl halide; also the π orbital system of the aromatic nucleus is likely to repel an approaching nucleophile. If, however, the aromatic nucleus is substituted by electron-withdrawing groups, particularly if they are o- and/or p- to the carbon atom carrying the halogen, then attack by a nucleophile is found to be greatly facilitated, and picryl chloride (17) is found to be almost as reactive as an acid chloride towards nucleophiles:

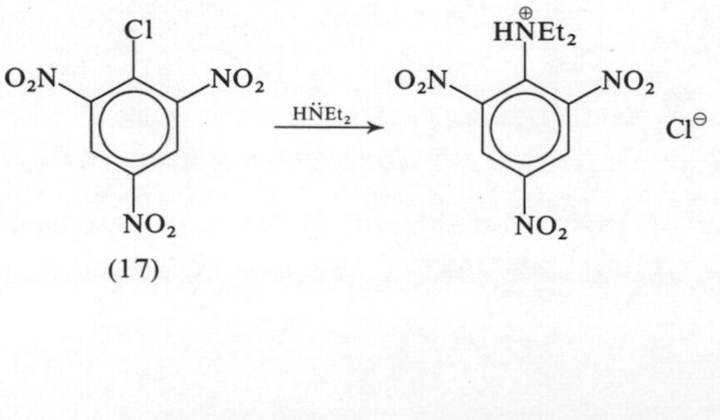

(17)

Such reactions of activated aryl halides are found to obey the generalized rate law

$$\text{Rate} = k\,[\text{ArCl}]\,[\text{Nucleophile}]$$

which is reminiscent of the S_N2 pathway for nucleophilic displacement on simple alkyl halides (*cf.* p. 23). There are, however, clear differences between the two in that here an sp^2, as opposed to an sp^3, hybridized carbon is attacked, and also attack from the back along the line of the C—Hal bond, characteristic of S_N2, cannot here apply as the aromatic nucleus, and its electron cloud, prevents such a direction of approach:

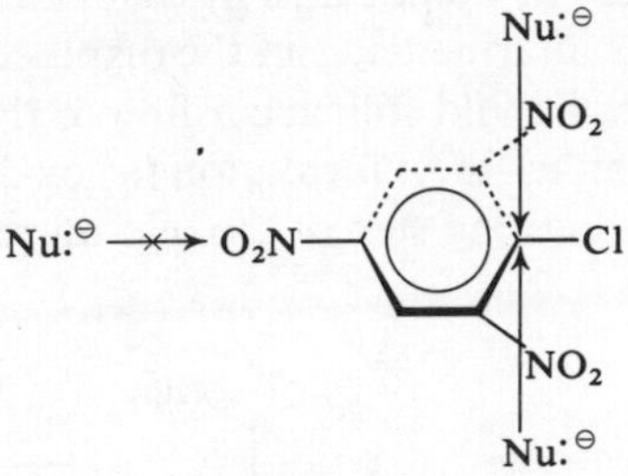

Attack must thus occur from the side (from above or below the plane of the paper in this formulation) more or less at right angles to the plane of the aromatic ring.

The fact that it is an sp^2 hybridized carbon that is here attacked allows of the complete formation of the new (Nu—C) bond before the old (C—Cl) bond need begin to be broken—an option not open to an sp^3 hybridized carbon. The reaction pathway could thus involve the formation, and subsequent decomposition, of an intermediate such as (18) as an alternative to proceeding *via* a single transition state in a one-step reaction (*cf.* S_N2):

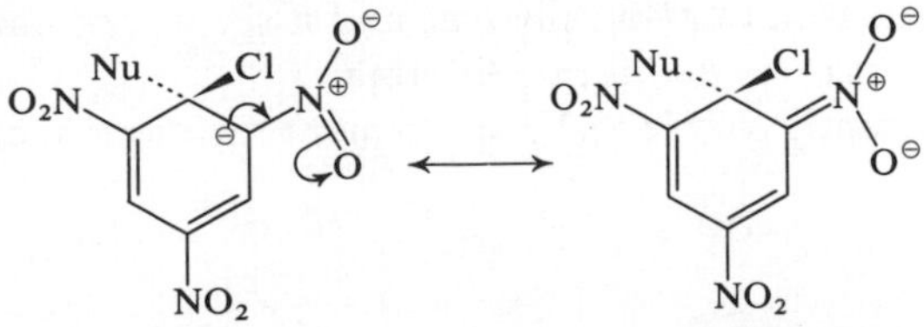

(18)   (18a)

Furthermore, electron-withdrawing groups in the 2-, and 4-positions are particularly well suited to stabilize such an intermediate by delocalization of its negative charge (18 ←→ 18a). Species such as (19), generally analogous to (18), have actually been isolated

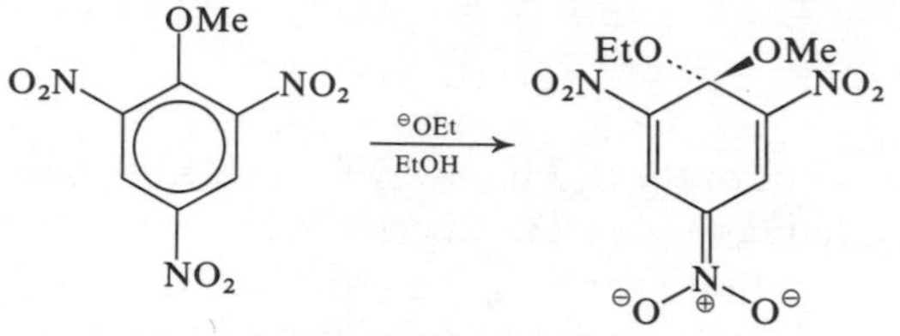

(19)

but no species exactly resembling (18), i.e. containing halide, has been isolated though detailed kinetic and other evidence strongly supports their involvement as intermediates in the displacement reaction. Given that such is the case, it could still, according to the generalized rate law specified above, be either their formation (*a*) or decomposition (*b*) that constituted the rate-limiting step of the overall reaction:

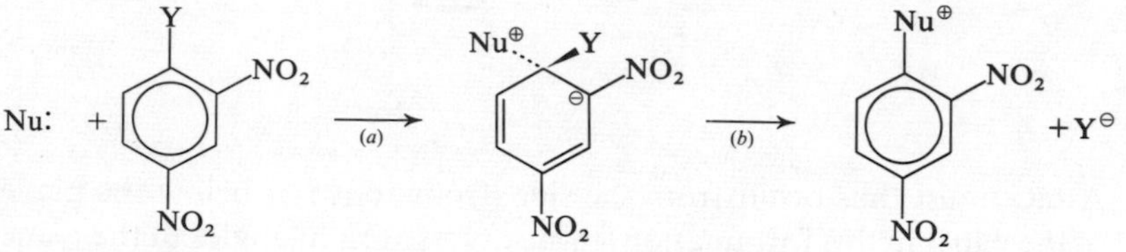

(20)

It has been shown for the attack of piperidine (Nu = $\bigcirc$) on a

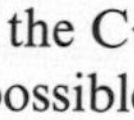

number of variants of (20), in which Y was among other things Cl, Br and I, that the overall rate of reaction did not vary significantly from one to another. This indicates that breaking of the C—Y bond was not, in these cases, involved to any major extent in the rate-limiting step of the overall reaction, which must thus be (*a*), formation of the C—Nu bond. But this is not universally true and it seems not impossible that in some cases it is (*b*) that is rate-limiting.

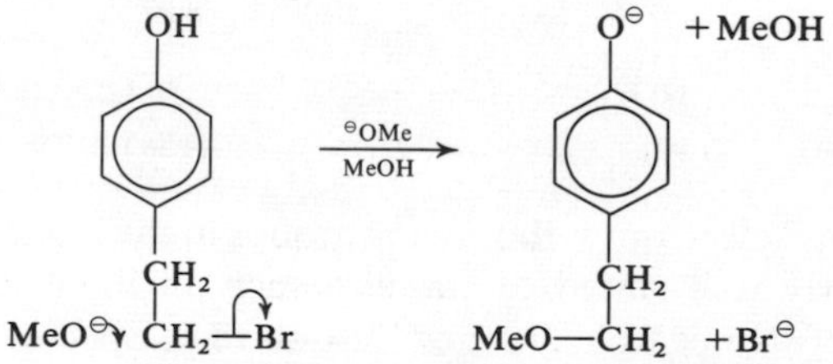

Even apparently simple S_N2 displacement reactions such as do not necessarily proceed in exactly the manner expected. Working under very carefully controlled conditions it is possible in the above reaction to isolate the species (21)

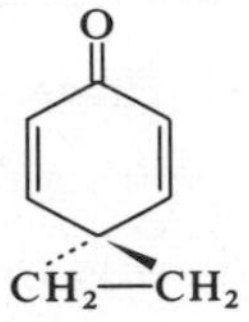

(21)

and to show that it may be decomposed, under the reaction conditions, to yield the normal products at a rate at least as fast as the overall reaction. The actual reaction pathway is probably:

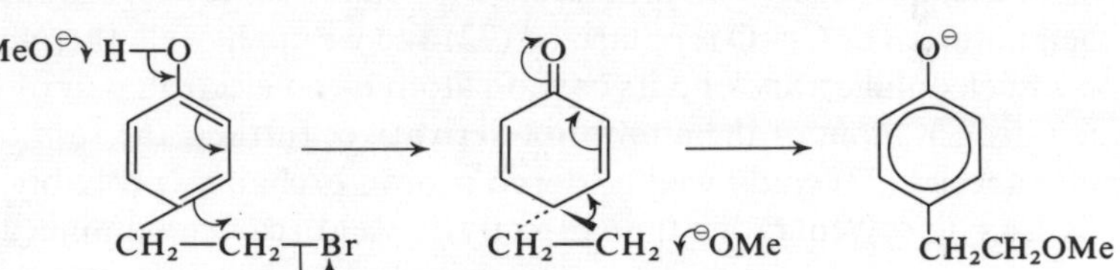

(21)

DETECTION OF INTERMEDIATES

Actual isolation of intermediates remains a pretty rare occurrence and our knowledge of reaction pathways would be slim indeed if we had to rely on it as our main source of information. Fortunately the development of physical, particularly spectroscopic, techniques in relatively recent times has provided us with a great deal of detailed and unequivocal information about unstable and/or fugitive species generated in the course of reactions.

(a) Spectroscopic. (i) Infra red:

Apart from the direct identification of species present in a reaction mixture, intermediates may often be inferred from the associated disappearance/appearance of particular groupings, linkages etc. in reactants and products, and some suggestions may thus be made about their likely structure. Thus in the reaction of carbonyl compounds with ammonia derivatives such as hydroxylamine and semicarbazide at neutral pH

$$\diagdown C = \overset{..}{N} \overset{\cdot}{} \quad + H_2O$$
$$\text{OH}$$

$$\diagup C = O$$

$$\overset{NH_2OH}{\nearrow}$$

$$\overset{NH_2NHCONH_2}{\searrow}$$

$$\diagdown C = \overset{..}{N} \overset{\cdot}{} \quad + H_2O$$
$$\text{NHCONH}_2$$

it was noticed (p. 10) that i.r. spectroscopy showed the C=O absorption (characteristic of the starting material) disappearing very rapidly (<2 min.), and the C=N absorption (characteristic of the product) appearing very much more slowly. The original carbonyl compound thus cannot be converted *directly* into oxime or semicarbazone, clearly one or more intermediate species must be gone through on the way. General chemical insight might enable us to make some suggestions as to their nature. The C=O is polarized (22) and we might well, therefore, expect nucleophilic attack on its carbon atom by an electron pair of the basic nitrogen atom of the ammonia derivative; further, the resultant dipolar species (23) could well undergo proton exchange (probably *via* the hydroxylic solvent rather than directly) to yield a carbinolamine (24):

$$\begin{array}{ccc}
\overset{\overset{\displaystyle O^{\delta-}}{\|}}{-\underset{|}{\overset{|}{C}}{}^{\delta+}} \quad :\!\overset{H}{\underset{H}{\overset{|}{N}}}\!-OH & \overset{\overset{\displaystyle O^{\ominus}}{|}}{-\underset{|}{\overset{|}{C}}}\!-\overset{H}{\underset{H}{\overset{|}{N}{}^{\oplus}}}\!-OH & \overset{\overset{\displaystyle OH}{|}}{-\underset{|}{\overset{|}{C}}}\!-\overset{H}{\underset{H}{\overset{|}{N}}}\!-OH \\
(22) & \xrightarrow[(a)]{} \quad (23) & \xrightarrow[(b)]{} \quad (24)
\end{array}$$

That such a sequence is not inherently unlikely is demonstrated by the fact that species such as (24a) can actually be isolated from the attack of ammonia derivatives on some carbonyl compounds, e.g. trichloroacetaldehyde (chloral):

$$Cl_3C - \overset{\displaystyle O}{\overset{\diagup\!\!\!/}{C}}\diagdown_H \;\; + NH_2OH \longrightarrow Cl_3C - \overset{\displaystyle OH}{\underset{\displaystyle NHOH}{\overset{\diagup}{\underset{\diagdown}{C}}}}\!\!-H$$
$$(24a)$$

To complete the formation of product from intermediate (24), dehydration is necessary: a process that is often acid-catalysed (*cf.* dehydration of alcohols → alkenes). We could envisage this happening *via* the sequence:

$$\begin{array}{ccc}
\overset{\overset{\displaystyle \ddot{O}H}{|}}{-\underset{|}{\overset{|}{C}}}\!-\overset{H}{\underset{H}{\overset{|}{N}}}\!-OH & \overset{\overset{\displaystyle \overset{\oplus}{\ddot{O}}H}{|}}{-\underset{|}{\overset{|}{C}}}\!-\overset{H}{\underset{H}{\overset{|}{N}}}\!-OH & \overset{\displaystyle H_2O}{-\underset{\diagdown\!OH}{\overset{|}{C}}} = \overset{..}{N}\overset{\cdot}{} \\
(24) & \underset{H_2O:}{\overset{H^{\oplus}}{\underset{(c)}{\rightleftharpoons}}} & \xrightarrow[(d)]{} \quad H_3O^{\oplus}
\end{array}$$

So far as the overall process at neutral pH is concerned, we know from the spectroscopic evidence that (*a*) is fast (disappearance of C=O); (*b*), a proton exchange, is also likely to be fast, and the slow part of the reaction is thus the acid-catalysed dehydration by which the product is produced (slow appearance of C=N absorption), i.e. (*c*) + (*d*).

Some test of this proposed pathway should be provided by making the reaction solution more acidic, for the slow (acid-catalysed) dehydration step should then be speeded up. This can indeed be observed spectroscopically and, in addition, the previously rapid disappearance of C=O, (*a*), is now found to be slowed down; it may even become rate-limiting. This slowing is due to increasing protonation of the basic nitrogen atom of $:NH_2OH$ as the acidity of the solution is increased:

$$H_3O^\oplus + :NH_2OH \rightleftharpoons H_2O: + H\overset{\oplus}{:N}H_2OH$$

The protonated species cannot function as a nucleophile and so does not attack C=O; the concentration of the effective nucleophile, $:NH_2OH$, is thus progressively reduced. All this adds up to the fact that the overall reaction is found to have an optimum pH at which its rate is a maximum, the exact pH value at which this occurs depending on the nature of both the carbonyl compound and the attacking nucleophile (*cf.* p. 176).

A somewhat similar situation is encountered in nucleophilic attack on the carbonyl group in carboxylic acid derivatives in, for example, transesterification reactions:

$$\overset{\displaystyle OEt}{\underset{\displaystyle R-C=O}{|}} + {}^\ominus OMe \rightleftharpoons \overset{\displaystyle OMe}{\underset{\displaystyle R-C=O}{|}} + {}^\ominus OEt$$

These reactions, given the second order rate law that they exhibit, could proceed either *via* a one-step displacement

$$R-\overset{\overset{\displaystyle \curvearrowright OEt}{|}}{\underset{\underset{\displaystyle {}^\ominus OMe}{}}{C}}=O \longrightarrow R-\overset{\overset{\displaystyle {}^\ominus OEt}{|}}{\underset{\underset{\displaystyle OMe}{|}}{C}}=O$$

or *via* a two-step process involving a tetrahedral (sp^3) intermediate (25):

$$R-\overset{\overset{\displaystyle OEt}{|}}{\underset{\underset{\displaystyle {}^\ominus OMe}{}}{C}}=O \longrightarrow R-\overset{\overset{\displaystyle \curvearrowright OEt}{|}}{\underset{\underset{\displaystyle OMe}{|}}{C}}-O^\ominus \longrightarrow R-\overset{\overset{\displaystyle {}^\ominus OEt}{|}}{\underset{\underset{\displaystyle OMe}{|}}{C}}=O$$

(25)

No intermediates such as (25) have been isolated nor, in such simple examples, have any been directly detected.

With the chloral adduct (24a) in mind, however, we might most hopefully look for a rather more stable carboxylate ester adduct among examples in which R was a powerfully electron-withdrawing-group. Observation of the i.r. spectrum of CF_3CO_2Et in solution in dibutyl ether demonstrates the very rapid (and complete) disappearance of the $C=O$ absorption band (1790 cm^{-1}) when NaOMe or NaOEt is added; H_2O and EtOH do not cause the band to disappear, nor does NaOMe in methanol ($^{\ominus}OMe$ is then a much poorer nucleophile because of its solvation by MeOH molecules). The CF_3CO_2Et may be recovered quantitatively by addition of gaseous hydrogen chloride. Similar disappearance of the $C=O$ absorption band (1750 cm^{-1}) also occurs with CF_3CONH_2, but only to the extent of ≈ 50 per cent. This latter observation is significant in that if in so highly electron-withdrawing a compound as this the $C=O$ is only 50 per cent transformed into adduct, it is not surprising that no significant concentration of adduct can be observed with CH_3CO_2Et or CH_3CONH_2. The formation of (25a)

$$
\begin{array}{ccc}
\overset{\displaystyle OEt}{\underset{\displaystyle \underset{\ominus}{C}OMe}{\overset{\displaystyle |}{CF_3-C\overset{\curvearrowright}{=}O}}} & \rightleftharpoons & \overset{\displaystyle OEt}{\underset{\displaystyle \underset{\displaystyle OMe}{|}}{\overset{\displaystyle |}{CF_3-C-O^{\ominus}}}} \\
& & (25a)
\end{array}
$$

does, however, suggest strongly that tetrahedral intermediates such as (24) are formed, and decomposed, during normal nucleophilic attack on carboxylic acid derivatives, though their equilibrium concentration is no doubt very small.

(ii) Nuclear magnetic resonance:

In the addition of a halogen, e.g. bromine, under polar conditions to an alkene

$$CH_2=CH_2 + Br_2 \longrightarrow BrCH_2-CH_2Br$$

it has long been known that carrying out the addition in the presence of an added nucleophile, $Y^{\ominus}$ (e.g. $Cl^{\ominus}$, $NO_3{}^{\ominus}$ etc.), though it does not affect the overall rate of the reaction, results in the formation of products other than the expected dibromide:

$$CH_2=CH_2 \xrightarrow[Y^{\ominus}]{Br_2} BrCH_2-CH_2Br + BrCH_2-CH_2Y$$

These products are formed at a rate faster than the reaction of $Y^{\ominus}$, under these conditions, on first-formed $BrCH_2CH_2Br$, and they must thus be formed in parallel with the latter, presumably from a common intermediate; bromine addition thus cannot take place by a pathway such as:

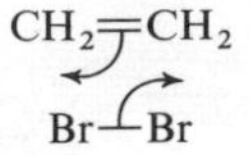

$$CH_2\!=\!CH_2$$
$$Br\!-\!Br$$

Any valid reaction pathway must allow for formation of the mixed product and one that has been suggested involves mutual polarization of the reactants (27), bonding to halogen to form an intermediate such as (26), and final formation of the end-product by attack of $Br^{\ominus}$ (or $Y^{\ominus}$) on (26):

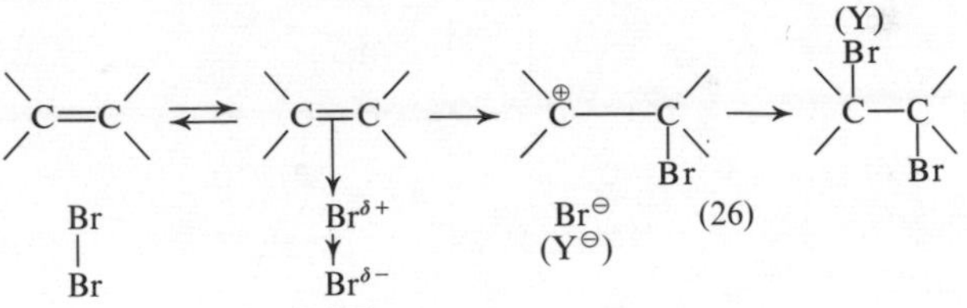

This explains why, in the presence of Y°, $BrCH_2CH_2Y$ is formed as well as $BrCH_2CH_2Br$, for $Br^{\ominus}$ is in no specially privileged position compared with $Y^{\ominus}$ so far as attack on (26) is concerned. It does not, however, adequately explain the fact that such polar additions of halogen are found, in simple cases at least, to proceed almost entirely TRANS (*cf.* p. 140); thus maleic acid (*cis*-$HO_2CCH\!=\!CHCO_2H$) gives only $(\pm)$1,2-dibromosuccinic acid ($HO_2CCHBrCHBrCO_2H$: work this one out for yourself) and cyclopentene (28) only the *trans*-dibromide (29):

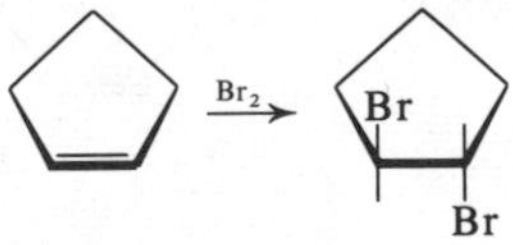

To account for this stereospecificity of addition it was suggested that the reaction proceeded not through a carbonium ion intermediate such as (26), but *via* a cyclic halonium (e.g. bromonium) ion such as (30); this could arise either from participation of an electron pair of the bromine atom in first-formed (26)

$$\overset{\diagdown}{\diagup}C\overset{\oplus}{-}C\overset{\diagup}{\diagdown} \quad \longleftrightarrow \quad \overset{\diagdown}{\diagup}C\overset{\diagup}{-}C\overset{\diagup}{\diagdown}$$

with $:Br$ (26) and $\underset{\oplus}{Br}$ (30)

or directly:

$$\overset{\diagdown}{\diagup}C\overset{\diagup}{=}C\overset{\diagup}{\diagdown} \quad \longrightarrow \quad \overset{\diagdown}{\diagup}C\overset{\diagup}{-}C\overset{\diagup}{\diagdown}$$

(27) with $Br^{\delta+}$, $Br^{\delta-}$ and (30) with $\underset{\oplus}{Br}$, $Br^{\ominus}$

Subsequent attack by an anion (e.g. $Br^{\ominus}$, $Y^{\ominus}$) would then take place only from the side of the bromonium ion remote from its bulky bromine atom

$$\overset{\diagdown}{\diagup}C\overset{\diagup}{-}C\overset{\diagup}{\diagdown} \quad \longrightarrow \quad \overset{\diagdown}{\diagup}C\overset{\diagup}{-}C\overset{\diagup}{\diagdown}$$

(30)

and the observed stereospecific addition would thus be explained. The cyclic bromonium ion was first proposed on an *ad hoc* basis specifically to explain the observed stereochemistry of this addition reaction; subsequently there has been inferential evidence from several other areas (*cf.* p. 135) supporting its existence, but no 'direct' detection in any sense.

Recently, however, 1,2-dibromides (and other 1,2-dihalides) such as (31) have been converted into ion pairs in 'super acid' solutions (SbF_5 in liquid SO_2 at $-60°$):

$$\underset{(31)}{\overset{CH_3}{\underset{CH_3}{\diagdown}}C\overset{Br}{\underset{Br}{-}}C\overset{CH_3}{\underset{CH_3}{\diagup}}} \quad \xrightarrow[\substack{\text{liq. } SO_2 \\ -60°}]{SbF_5} \quad \underset{(32)}{\overset{CH_3}{\underset{CH_3}{\diagdown}}C\overset{\oplus}{\underset{Br}{-}}C\overset{CH_3}{\underset{CH_3}{\diagup}}} \quad SbF_5Br^{\ominus}$$

The one signal ($\tau 8.0$) proton n.m.r. spectrum of the starting material (31) is shifted downfield ($\tau 7.1$) in the resultant ion, as would be expected, but the latter is found to give rise not to the two signal spectrum that (32) would require (arising from the different signals from two non-equivalent pairs of methyl groups), but to one proton signal only: all four methyl groups must thus be equivalent. This could perhaps be explained by a rapid, equilibrating shift

$$
\underset{(32)}{
\begin{array}{c}
CH_3 \diagdown \qquad \diagup CH_3 \\
C\!-\!\overset{\oplus}{C} \\
CH_3 \diagup \ \ Br \qquad \diagdown CH_3
\end{array}
}
\quad \rightleftharpoons \quad
\underset{(32a)}{
\begin{array}{c}
CH_3 \diagdown \qquad \diagup CH_3 \\
\overset{\oplus}{C}\!-\!C \\
CH_3 \diagup \qquad Br\ CH_3
\end{array}
}
$$

which, if fast enough, could lead to a single, time-averaged signal. No significant difference in n.m.r. spectrum is noticed on lowering the temperature as far down as $-120°$, however, by which time one would certainly expect the $(32 \rightleftharpoons 32a)$ shift to be so slowed as to yield two signals. It thus seems virtually certain that by ion pair formation from (31) we are in fact observing, directly, the generation of a bromonium ion (30a);

$$
\underset{(30a)}{
\begin{array}{c}
CH_3 \diagdown \qquad \diagup CH_3 \\
C\!-\!\!-\!\!C \\
CH_3 \diagup \ \ \underset{\oplus}{Br} \diagdown CH_3
\end{array}
}
$$

Its generation under such widely different conditions does not of course establish the participation of (30a) as an intermediate in the addition of bromine to alkenes, but it does at least put an end to its essentially *ad hoc* participation in the latter, and greatly increases its inherent credibility.

N.m.r. spectroscopy has been particularly successful in the direct detection of carbonium ions. Thus dissolving the tertiary halide, 2-fluoro-2-methylpropane (33) in an excess of antimony pentafluoride results in a shift of proton resonance downfield and the appearance of a single signal ($\tau 5.3$) arising from the nine equivalent protons in the tertiary (2-methylpropyl) cation (34):

$$
\underset{(33)}{(CH_3)_3C\!-\!F} + SbF_5 \rightleftharpoons \underset{(34)}{(CH_3)_3C^{\oplus}SbF_6^{\ominus}}
$$

Use of the even stronger acid system, FSO_3H/SbF_5, results in the ionization of alcohols:

$$
(CH_3)_3C\!-\!\underset{\cdot\cdot}{O}H \xrightleftharpoons{FSO_3H} (CH_3)_3C\!-\!\overset{\oplus}{\underset{H}{O}}H \underset{/SbF_5}{\xrightleftharpoons{FSO_3H}} \underset{(34)}{(CH_3)_3C^{\oplus}SbF_5FSO_3^{\ominus} + H_3O^{\oplus}}
$$

It is significant that the n.m.r. spectrum of primary and secondary alcohols in FSO_3H/SbF_5 is often time-variable indicating rearrangement of the first-formed cation,

$$CH_3CH_2CH_2CH_2OH \xrightarrow[/SbF_5]{FSO_3H} CH_3CH_2CH_2CH_2{}^{\oplus}SbF_5FSO_3{}^{\ominus} \rightsquigarrow (CH_3)_3C^{\oplus}SbF_5FSO_3{}^{\ominus}$$
(34)

ending up, where this is feasible, with the particularly stable tertiary cation (34). The stability of this species, under these conditions, is indeed so great that it is possible to abstract hydride ion from an alkane in order to obtain it:

$$(CH_3)_3C-H \xrightleftharpoons[/SbF_5]{FSO_3H} (CH_3)_3C^{\oplus}SbF_5FSO_3{}^{\ominus} + H_2$$
(34)

Here again, while not actually identifying intermediates in particular reactions, we are nevertheless directly establishing the credibility of carbonium ions as species with a real existence: they thus become that much more plausible as potential intermediates.

(iii) Electron spin resonance:

We have referred above (p. 73) to several of the characteristics that identify a radical reaction—initiation by light or added radicals, termination by added inhibitors etc.—but the most direct method of identification makes use of the fact that radicals contain an unpaired electron, i.e. that they are *paramagnetic*. The spin of such an unpaired electron gives rise to a magnetic moment which is able to orient itself with or against an applied magnetic field; these two states correspond to different energy levels and transitions between them lead to characteristic absorption lines in the microwave region of the spectrum. The exploitation of these signals is known as electron spin (or paramagnetic) resonance spectroscopy; it is capable of detecting radical intermediates in concentrations as small as 10^{-12} molar, and of providing considerable information about their structure.

The latter possibility arises from the fact that the observed signals have a fine structure due to interaction of the unpaired electron with nuclei that have a magnetic moment e.g. ^1H and ^{13}C, but not ^{12}C. Thus while study of the e.s.r. spectrum of methyl radicals does not unfortunately allow us to distinguish between the alternative structures (35a) and (35(b)

(35a)   (35b)

investigation of $H_3{}^{13}C\cdot$ shows that it, and by implication $H_3{}^{12}C\cdot$, is in fact planar (35a) at least to within 5°. Planarity does not, however, seem to be an essential characteristic, and non-planar radicals are not uncommon.

Some limitation attended the early use of e.s.r. spectroscopy in that to make measurements it was often necessary to trap and preserve radicals, because of their otherwise low concentration, in the holes of solid matrices, e.g. of frozen krypton, neon etc. This clearly inhibits the study of reaction intermediates, but measurements have now been increasingly extended to work in solution as, for example, in hydrogen abstraction from molecules by hydroxyl radicals generated for the purpose. Thus attack on cycloheptatriene (36) was found to result in a spectrum much less complex than would be exhibited by the expected (37):

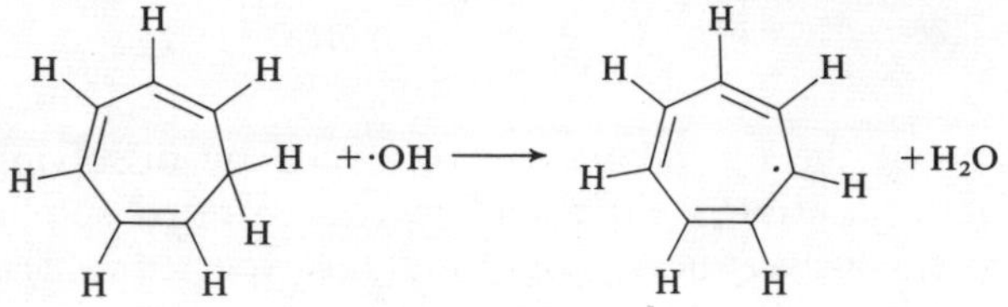

(36)   (37)

It actually comprised eight equally spaced lines indicating a symmetrical structure for the cycloheptatrienyl radical in which the spin of the electron was interacting with seven equivalent protons. The structure of the radical must thus be (38)

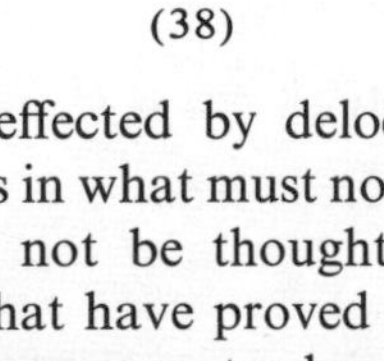

(38)

in which stabilization is effected by delocalization of the electron over all seven carbon atoms in what must now be a planar species.

These examples should not be thought to exhaust the types of spectroscopic techniques that have proved useful in the investigation of intermediates. Thus Raman spectra have been used (see below, p. 91), and the use of u.v. spectra in investigating the decarboxylation of β-keto acids is also referred to (p. 102). Mass spectra have the limitation of operating in the vapour phase under greatly reduced pressure, but they too have been successfully and ingeniously employed in the investigation of intermediates; an example, the detection of benzynes, is discussed subsequently (p. 100).

(b) Other physical measurements

One physical measurement that has yielded useful inferential information about the nature of reacting species has been the depression of freezing point of concentrated sulphuric acid produced by their dissolution in it. An early example of this arose from the role played by the sulphuric acid component of nitrating mixture in aromatic nitration:

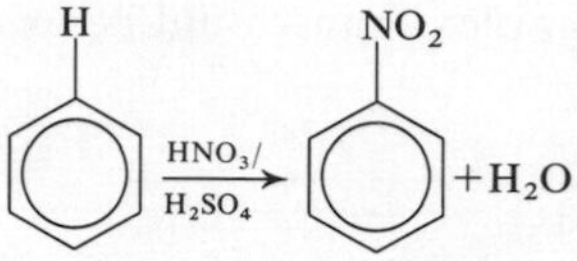

Investigation of the f.p. depression produced by dissolving known amounts of concentrated nitric acid in concentrated sulphuric acid showed depressions very nearly four times the expected values, indicating the formation of four species. The fact that sulphuric is a stronger acid than nitric suggests that the initial interaction may be reversible protonation of the latter by the former:

$$H_2SO_4 + H\underset{..}{O}-NO_2 \rightleftharpoons HSO_4^\ominus + H\overset{\oplus}{\underset{..}{O}}-NO_2$$
$$\underset{H}{}$$
$$(39)$$

This would, however, still yield only two species per HNO_3 added, but the not unreasonable fission of (39) could yield (40) and water, that would, in turn, immediately be protonated:

$$H\overset{\oplus}{O}\overset{\frown}{}NO_2 \rightleftharpoons H_2O + {}^\oplus NO_2$$
$$\underset{H}{}$$
$$\quad(39) \qquad\qquad (40)$$

$$H_2O + H_2SO_4 \rightleftharpoons H_3O^\oplus + HSO_4^\ominus$$

Summation of these several reactions would lead to the stoichiometry

$$H\underset{..}{O}-NO_2 + 2H_2SO_4 \rightleftharpoons {}^\oplus NO_2 + H_3O^\oplus + 2HSO_4^\ominus$$
$$(40)$$

in which four species are indeed produced per molecule of HNO_3 added; this rationalization has the added virtue that (40) would be expected to be a powerful electrophile and, in this context, a powerful nitrating agent. Some confirmation is provided by the fact that a nitric acid/perchloric acid mixture is also a powerful nitrating agent, i.e. sulphuric acid is not an essential constituent of nitrating mixture (it is however, commonly used because it is a good solvent for aromatic species): all that is required is an acid sufficiently stronger than nitric acid to be able to encompass the latter's protonation.

Study of the Raman spectra (examination of the radiation scattered by a pure liquid or a concentrated solution, when irradiated with a narrow band of radiation) of solutions of nitric acid in sulphuric acid reveal the presence of an absorption band at $1400 \, \text{cm}^{-1}$ which can, theory dictates, arise only from a diatomic, or a symmetrical, linear triatomic, molecule. It is difficult to visualize the formation of any diatomic species, and the same absorption is also exhibited in mixtures of nitric and perchloric acids, in stable salts such as $^{\oplus}NO_2 \, BF_4^{\ominus}$, $^{\oplus}NO_2 \, ClO_4^{\ominus}$ etc.—some of whose structures have been established unequivocally by X-ray crystallography—and also, very faintly, in nitric acid itself (arising from a very small amount of autoprotonation). The case for the absorption being due to the *nitronium ion* (40)

$$O{=}\overset{\oplus}{N}{=}O$$
(40)

and for this being present in nitrating mixture, is thus overwhelming, and there can be no sensible doubt that it is the effective nitrating agent therein.

A further example of obtaining information about an intermediate from f.p. depression measurements is discussed below (p. 118).

TRAPPING OF INTERMEDIATES

Trapping has the virtue of dealing directly with the reaction system, but there are limitations on its use in that the added species must not affect the course of the reaction being studied other than by 'cutting out' the intermediate. Clearly the added 'trap' must intercept the reaction pathway only after the formation of the relevant intermediate, and the best evidence that this condition is being met is provided by the absence of kinetic change (rate of disappearance of starting material) on adding the trap. It is obviously important also to resist designating an alternative end-product of the reaction as an intermediate just because it may happen to react with the added 'trapping' species.

(a) Radicals

We have already mentioned (p. 74) the effect of inhibitors in stopping radical chain reactions, and this can be looked upon as an example of trapping as the inhibitor substances owe their effectiveness to the ease with which they will react with, and hence scavenge, radical intermediates. One such inhibitor is hydroquinone (45) which is used as an *anti-oxidant*, i.e. it is added to substances to prevent their oxidation by the oxygen of the air (*autoxidation*)—the major source of decomposition of very many organic compounds. Such autoxidations are usually radical reactions, e.g. with benzaldehyde, the reaction being initiated by light, by traces of oxidative metal ions that are usually present, by peroxides (ROOR) or other radical sources:

$$
\underset{}{\overset{O}{\overset{\|}{Ph-C-H}}} + Fe^{3\oplus} \longrightarrow Fe^{2\oplus} + H^{\oplus} + \underset{(41)}{\overset{O}{\overset{\|}{Ph-C\cdot}}} \xrightarrow[\text{(from air)}]{\cdot O_2\cdot} \underset{(42)}{\overset{O}{\overset{\|}{Ph-C-O-O\cdot}}}
$$

$$
\underset{(41)}{\overset{O}{\overset{\|}{Ph-C\cdot}}} \quad + \quad \underset{(43)}{\overset{O}{\overset{\|}{Ph-C-O-OH}}}
$$

with $\downarrow$ PhCHO from (42)

$$
\underset{(43)}{\overset{O}{\overset{\|}{Ph-C-O-OH}}} + \underset{}{\overset{O}{\overset{\|}{Ph-C-H}}} \xrightarrow{H^{\oplus}} 2\underset{(44)}{\overset{O}{\overset{\|}{Ph-C-OH}}}
$$

Molecular oxygen can be looked upon as a rather unreactive diradical that is not normally itself capable of producing other radicals, e.g. by hydrogen abstraction etc., but that will react readily with any radicals present, e.g. the benzoyl radical (41), that have been produced in other ways. The resultant perbenzoate radical (42) is reactive, and can remove a hydrogen atom from a second molecule of benzaldehyde to produce perbenzoic acid (43) plus a further benzoyl radical (41) that, in turn, initiates another oxidation cycle: an oxidative chain reaction involving oxygen from the air has thus been set up with perbenzoic acid (43) as its product. The formation of the final end product, benzoic acid (44), results from attack of this perbenzoic acid on a further molecule of benzaldehyde in a non-chain process that is catalysed by acid; the rate of this latter step thus increases as product benzoic acid accumulates, i.e. it is auto-catalytic (*cf.* p. 29).

The effect of added hydroquinone (45) is to react preferentially with benzoyl radicals (41), thus bringing the oxidative chain reaction to a stop as the resultant radical (46) is not itself reactive enough to carry it on—probably because of the stabilizing effect of delocalization of the unpaired electron over the aromatic system. It can, however, scavenge a further chain-carrying radical to form quinone (47):

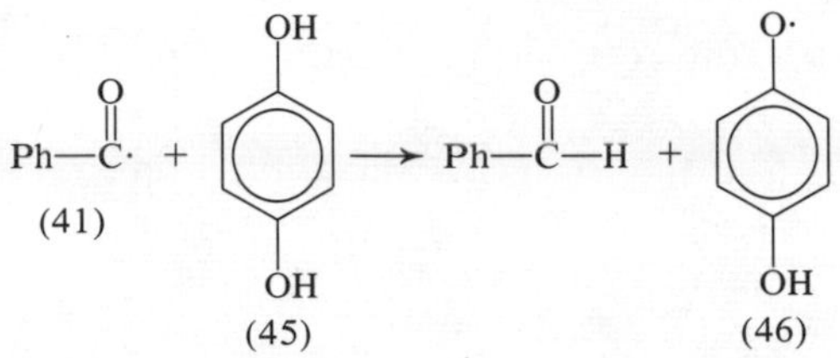

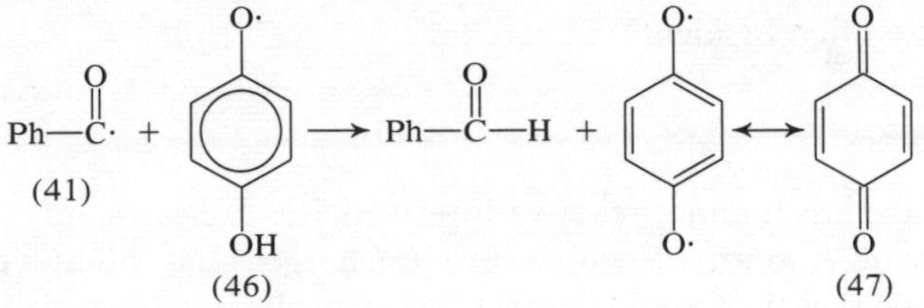

It may perhaps be objected that hydroquinone (45) can hardly be described as 'trapping' intermediates in that the specific (radical) intermediate is not identified by isolation of the 'trap' molecule with the intermediate attached to it. We have, however, identified the intermediates as radicals, and even the above objection can, in some cases, be overcome as quinone (47) also acts as a radical trap, but this time by incorporating them:

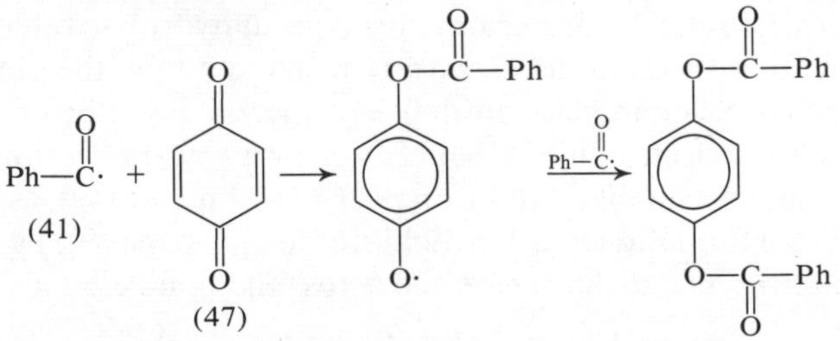

Another test that has been used to detect radical intermediates is their ability to initiate the polymerization of very readily polymerized species such as styrene (48):

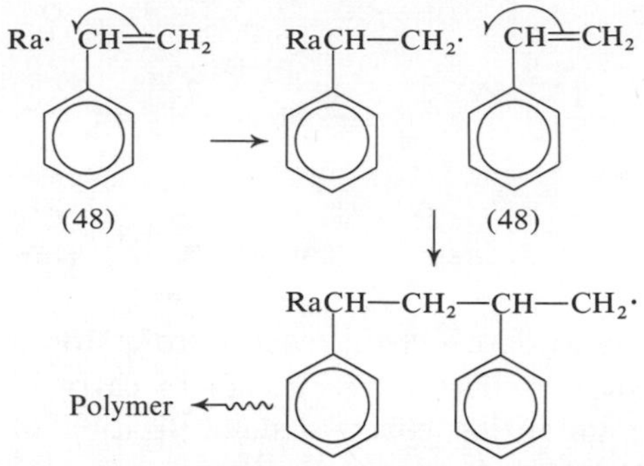

Ra· $\overset{\curvearrowright}{C}H\doublebond$CH$_2$ RaCH—CH$_2$· $^{\curvearrowright}$CH$\doublebond$CH$_2$

(48)   (48)

RaCH—CH$_2$—CH—CH$_2$·

Polymer ←~~~

Addition of monomeric styrene to a suspected radical reaction thus leads to polystyrene with the original radical intermediate incorporated as an end-group; in polymers of not too great chain length this end group can often be identified.

(b) Carbenes

Among the less familiar reactive intermediates are carbenes, R$_2\ddot{C}$, and nitrenes, R$\ddot{\ddot{N}}$, both of which are uncharged but highly electron-deficient in that the C and N atoms, respectively, have only six electrons in their outer shells. The most familiar examples of the former are dihalocarbenes such as CCl$_2$, which is believed to be formed during the hydrolysis of chloroform in the presence of strong bases:

$$\text{HO}^{\ominus} \; \text{H}\overset{\curvearrowleft}{—}\overset{..}{C}\text{Cl}_3 \underset{\text{fast}}{\overset{-\text{H}_2\text{O}}{\rightleftharpoons}} {}^{\ominus}\text{CCl}_3 \xrightarrow{\text{slow}} \text{Cl}^{\ominus} + \text{CCl}_2 \xrightarrow[\text{fast}]{\text{H}_2\text{O}} \text{CO} \overset{{}^{\ominus}\text{OH/H}_2\text{O}}{\leadsto} \text{HCO}_2{}^{\ominus}$$

The initial removal of proton is believed to be fast as in deuterated solvents, e.g. D$_2$O, the rate of hydrolysis is slow compared with the rate of incorporation of deuterium into as yet unhydrolysed chloroform; the formation of dichlorocarbene is believed to be the slow, rate-limiting step. Such an electron-deficient species might be expected to be a powerful electrophile, a belief borne out by the fact that intro-duction, under suitable conditions, of *cis*-2-butene (49) as a 'trap' results in the formation of a 1,1-dichlorocyclopropane (50) by addition of the electrophilic dichlorocarbene across the double bond:

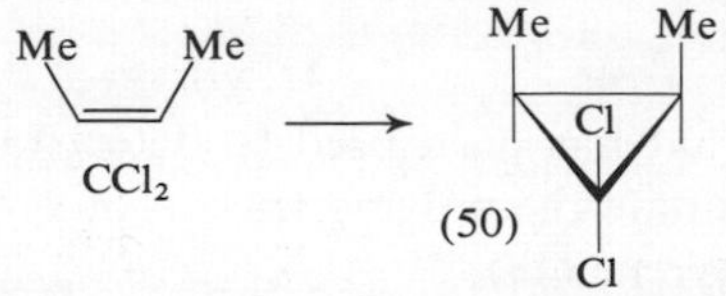

(49)   Me   Me
        \\===//   →   Me   Me
         CCl$_2$ \\   Cl   /
                                   ＼ /
                             (50)   V
                                     |
                                     Cl

A plausible pathway for the Reimer-Tiemann reaction of phenol, chloroform and a strong base to yield salicaldehyde (51) can be suggested that involves trapping of dichlorocarbene (generated from chloroform and base) as the salient step:

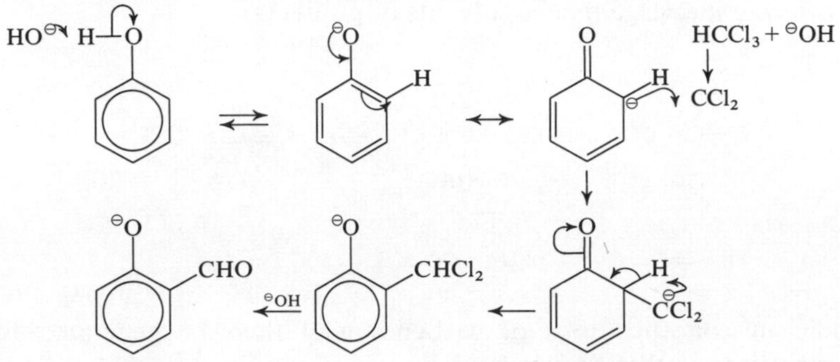

(51)   (52)

Even more cogent evidence in support of this as the key step is provided by the closely similar reaction with the anion of *p*-cresol (53) which yields, in addition to the expected *o*-aldehyde (54), the dichloro-compound (56);

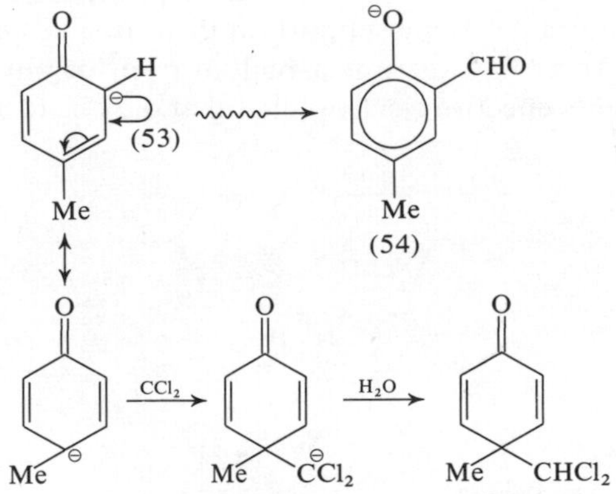

Here the carbene adduct (55) cannot eliminate a proton from the benzene nucleus to reform an aromatic structure—as did the corresponding adduct (52) from phenol—and therefore stabilizes itself by abstraction of a proton from the solvent to yield the end-product (56). The latter resists hydrolysis partly because it is somewhat insoluble in the aqueous alkali employed, and also because the halogens are in a hindered (neopentyl, *cf.* p. 115) environment that makes attack by hydroxyl ion very slow.

Dichloro-(and other)-carbenes are known not merely as intermediates in the hydrolysis of chloroform, but are produced in a number of other reactions particularly where these are carried out in the absence of hydroxylic species, either as solvents or products:

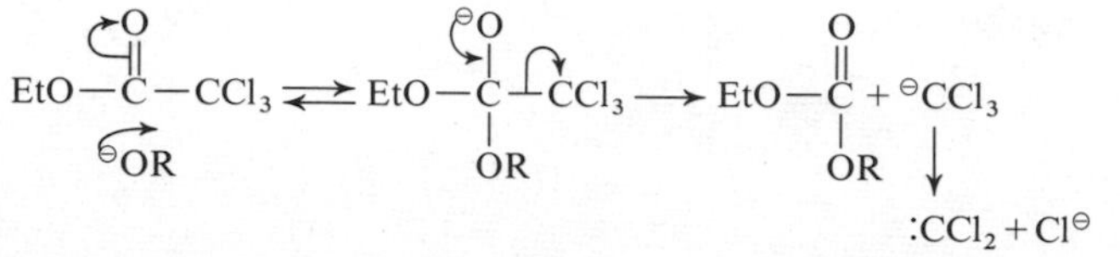

$$EtO-\overset{O}{\underset{OR}{C}}-CCl_3 \rightleftharpoons EtO-\overset{O}{\underset{OR}{C}}-CCl_3 \longrightarrow EtO-\overset{O}{\underset{OR}{C}} + {}^{\ominus}CCl_3$$

$$:CCl_2 + Cl^{\ominus}$$

$$Ph_2C{=}C{=}O \xrightarrow{\Delta} Ph_2C{:} + CO$$

Sufficient concentrations of carbenes may then be generated for preparative, synthetic procedures.

(c) Nitrenes

Nitrenes, $R\ddot{N}$, have been suggested as intermediates in a variety of reactions, among them the reduction of aromatic nitro, $ArNO_2$, and nitroso, $ArNO$, compounds with trivalent phosphorus compounds, e.g. triethyl phosphite. Some support for their involvement is afforded by choosing Ar so that there is a built-in trap for any intermediate nitrene—a highly effective electrophile—that may be formed:

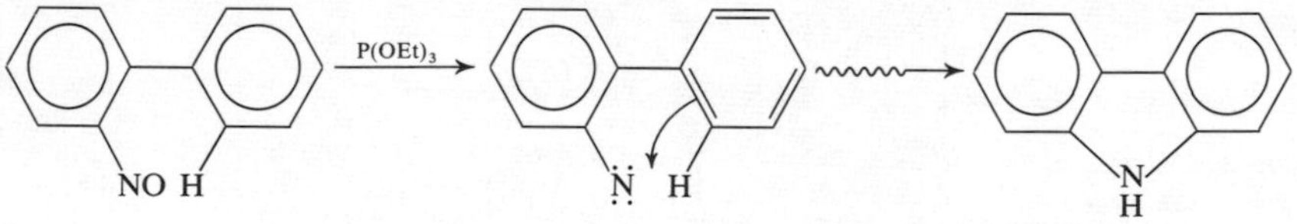

(57)

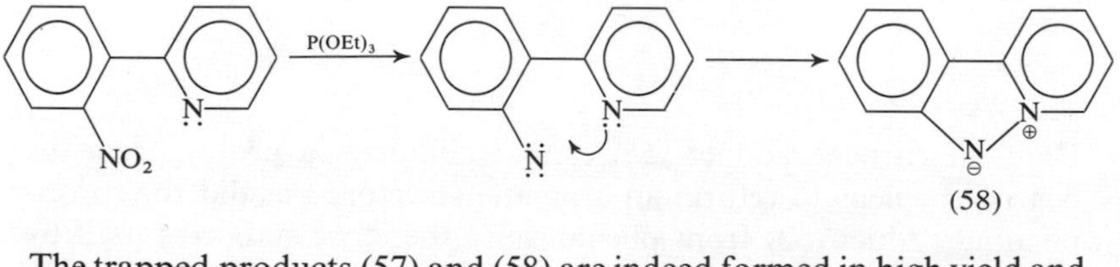

(58)

The trapped products (57) and (58) are indeed formed in high yield and though this does not identify $Ar\ddot{N}$ unequivocally as the intermediate involved, it is made the more likely by the fact that similar distributions of products are often obtained from photolysis of the corresponding azides (59),

$$Ar-\overset{\ominus}{\underset{(59)}{N}}-\overset{\oplus}{N}\equiv N \xrightarrow{h\nu} Ar-\ddot{N}+N\equiv N$$

whose decomposition to nitrenes is more inherently plausible, and also better documented.

(d) Benzynes

The observation that unactivated aryl halides such as 4-chlorotoluene (60) reacted with nucleophiles/bases to yield unexpected (62) as well as expected products (61)

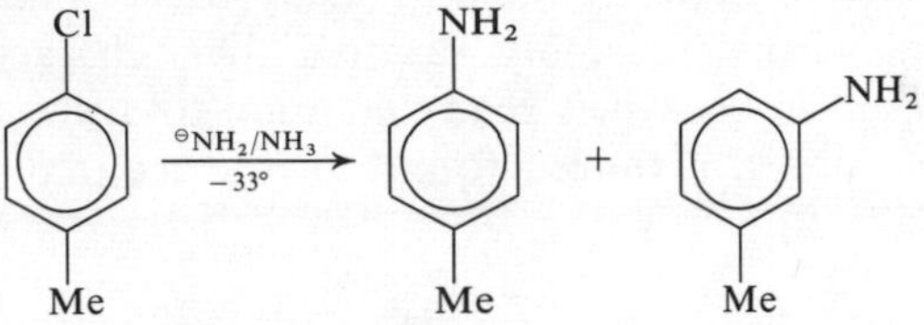

Cl
$\xrightarrow[-33°]{\ominus NH_2/NH_3}$
NH₂   +   NH₂

Me   Me   Me
(60)   (61)   (62)

demonstrated that the reaction could not be merely a simple displacement process, and led to the suggestion that benzyne intermediates (63) were involved (i.e. that 'displacement' is in fact elimination/addition):

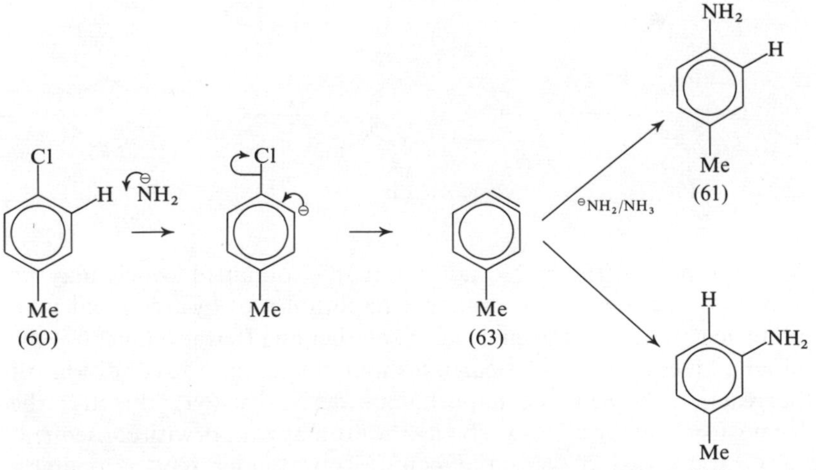

Cl
H ⁀NH₂
→
Cl
→

$\xrightarrow{\ominus NH_2/NH_3}$

NH₂
H
Me
(61)

H
NH₂
Me
(62)

Me   Me   Me
(60)   (63)

Such a suggestion receives some general support from the fact that no product is ever obtained in which the amino group has attacked the carbon atom *ortho* to that carrying the methyl group. It should be emphasized that benzyne is not an acetylene, whose linear sp^1 hybrid orbitals require an essentially linear arrangement of the two triply bonded carbons and the adjacent ones on each side, but a wholly aromatic system with two available sp^2 hybrid orbitals in the same plane:

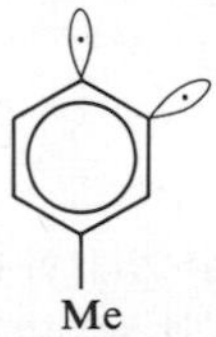

Me

Because of the way these two orbitals are arranged with respect to each other, overlap between them is likely to be minimal and the resulting bond weak, i.e. highly reactive. This character might be expected to result in benzyne's ready addition to dienes in the Diels–Alder reaction; and it is indeed found in practice that dienes act as excellent benzyne traps:

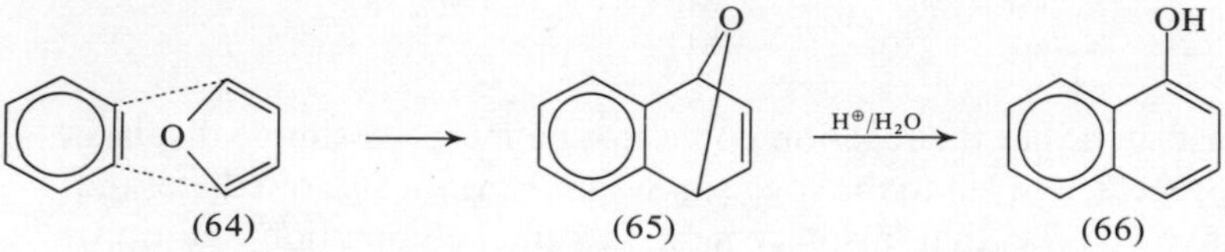

(64)   (65) $\xrightarrow{H^{\oplus}/H_2O}$ (66)

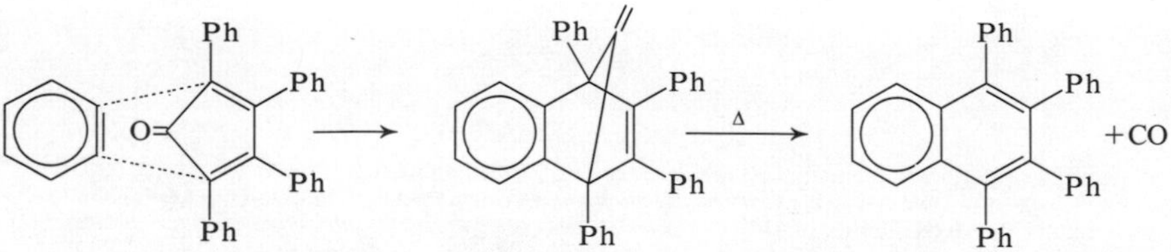

(67)   (68) $\xrightarrow{\Delta}$ (69) $+CO$

With furan (64), the stable adduct (65) is obtained which may be converted into the more familiar 1-naphthol (66) by treatment with dilute acid. With tetraphenylcyclopentadienone (tetracyclone, 67) the initial adduct (68) extrudes carbon monoxide under the conditions of the reaction to yield the naphthalene derivative (69), directly: the driving force for the extrusion being the aromatization, with consequent stabilization, that is thereby effected. Tetracyclone is a particularly useful trap as it is highly coloured (deep purple), while the adduct is not, so the trapping operation may be readily observed colorimetrically, or even by eye.

The ambient concentration of benzyne intermediates developed during the action of $^{\ominus}NH_2/NH_3$, $^{\ominus}OH$ etc. on unactivated aryl halides is small, as a high concentration of nucleophile is present to react with the benzyne as soon as it is formed. Much higher concentrations of benzynes may, however, be developed from suitable species that do not require nucleophiles for their initial decomposition, particularly if solvents of low polarity are used, e.g.:

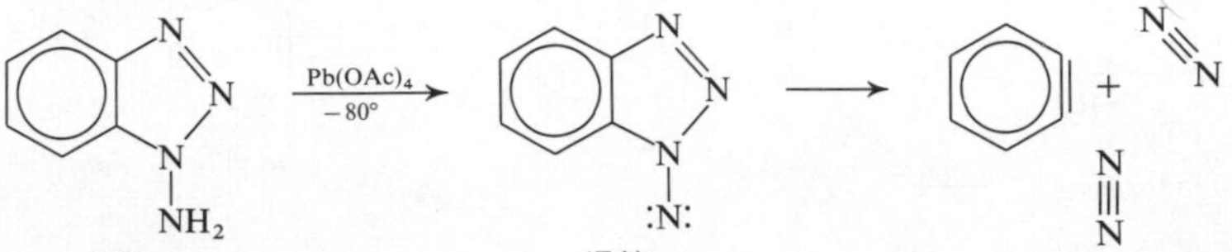

(70)   (71)

It is interesting that the initial intermediate from the oxidation of 1-aminobenztriazole (70) with lead tetraacetate at $-80°$ is the nitrene (71), which then loses two molecules of nitrogen (some driving force!) to yield the benzyne. Yields of the benzyne intermediate are high enough to allow of its being used synthetically, i.e. preparative trapping:

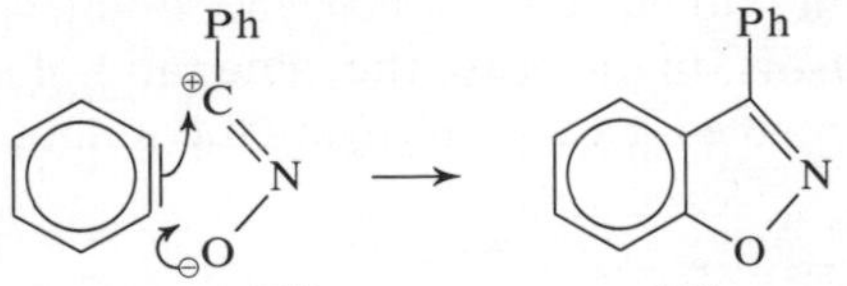

(72)   (73)

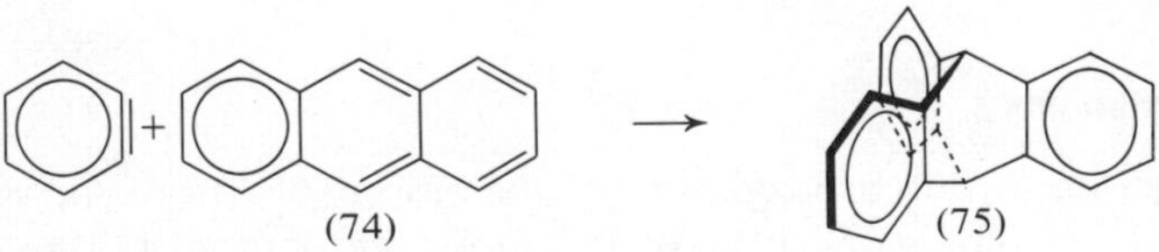

(74)   (75)

Thus benzonitrile oxide (72) acts as a 1,3-bipolar adduct to yield the benzisooxazole (73), while anthracene yields the interesting, cage-like hydrocarbon, triptycene (75). When no such preparative traps are added, benzyne molecules, which are present in high concentration, undergo self addition

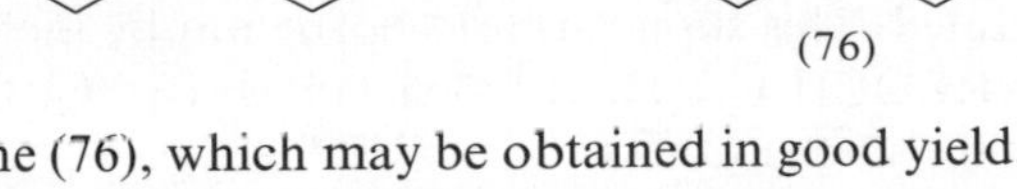

(76)

to biphenylene (76), which may be obtained in good yield.

A very neat detection of a benzyne intermediate by a physical method is provided when the zwitterion (78) from the diazotization of anthranilic acid (77) is decomposed in the heated inlet of a mass spectrometer. A very simple mass spectrum is developed with major lines corresponding to fragments of m/e 28, 44, and 76:

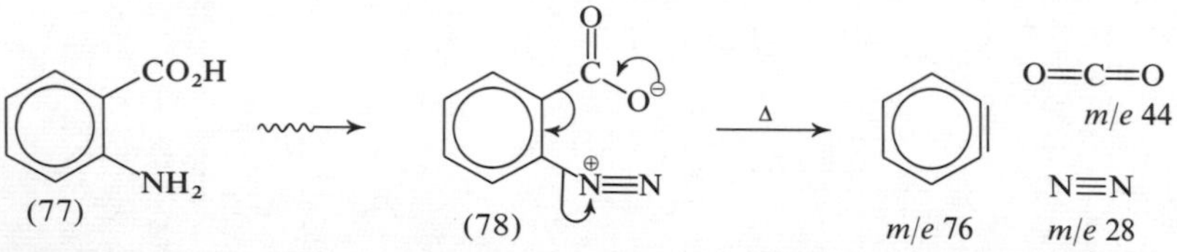

The mass spectrum is also found to be time-variable, the m/e 76 line declining rapidly and being replaced by one at m/e 152, due to progressive dimerization of benzyne to yield biphenylene (76).

The question of whether it is really the same type of intermediate, benzyne, that is generated from such diverse sources, in addition to the 'displacement' reactions of unactivated aryl halides, has been investigated by adding the same mixture of two different dienes—furan and cyclohexa-1,3-diene—to the reputed benzynes obtained from a number of different precursors. In each case the same ratio of adducts from the two dienes was obtained, clearly implying a common intermediate in all cases.

(e) Carbanions

The well-known decarboxylation of the anions of carboxylic acids is believed, in some cases at least, to involve carbanions (79) as intermediates:

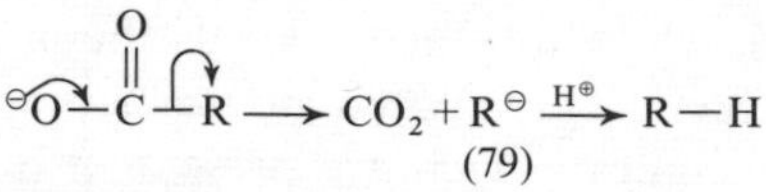

It would be expected that the presence of an electron-withdrawing group in R would stabilize a carbanion intermediate, and that decarboxylation might therefore proceed more readily in the salt of an acid so substituted. This supposition is borne out by the behaviour of β-keto acids, $RCOCH_2CO_2H$, trihaloacetates, e.g. Cl_3CCO_2H, and α-nitroacids (80), among others:

$$\overset{\ominus}{O}\overset{O}{\overset{\|}{-}}C\overset{\curvearrowright}{\overset{}{-}}CMe_2-\overset{\oplus}{N}=O \longrightarrow CO_2 + \left[\begin{array}{c} \overset{\ominus}{\overbrace{CMe_2}}\overset{\curvearrowright}{-}\overset{\oplus}{N}\overset{\curvearrowleft}{=}O \\ | \\ O^{\ominus} \\ \updownarrow \\ CMe_2=\overset{\oplus}{N}-O^{\ominus} \\ | \\ O^{\ominus} \end{array} \right] \overset{H^{\oplus}}{\longrightarrow} HCMe_2NO_2$$

$$\begin{array}{ccc} (80) & (81) & (82) \end{array}$$

Stabilized (delocalized) carbanions such as (81) may be trapped, and inferentially identified, by carrying out the decarboxylation in the presence of bromine; the overall rate of decarboxylation is unaffected, the end-product is, however, no longer the nitroalkane (82) but its brominated derivative (83). The normal end-product (82), and the starting material (80), are attacked only very slowly, if at all, by bromine under these conditions, whereas (83) is formed rapidly and must therefore be derived from a reactive precursor of (82):

$$\overset{\frown}{Br}\overset{}{-}Br \quad \overset{\frown}{\overset{\ominus}{C}Me_2NO_2} \longrightarrow Br^{\ominus} + BrCMe_2NO_2$$

$$\begin{array}{cc} (81) & (83) \end{array}$$

Kinetic studies on the decarboxylation of β-ketoacids show that the reaction generally involves the free acid as well as its anion,

$$\text{Rate} = k_1[RCOCH_2CO_2{}^{\ominus}] + k_2[RCOCH_2CO_2H]$$

the relative contribution of the two depending on the structure of the acid, and also on the conditions. The decomposition of the acid is believed to involve a six-membered transition state or intermediate in which the decarboxylation may, or may not, be a concerted process:

$$\begin{array}{ccc}
\overset{H}{\underset{O=C}{\overset{|}{O}}}\overset{}{\underset{}{C}}-CO_2H & \longrightarrow & O=C \quad C-CO_2H & \longrightarrow & CO_2 + H \quad C-CO_2H \\
\underset{Me \quad Me}{C} & & \underset{Me \quad Me}{C} & & \underset{Me \quad Me}{C} \\
(84) & & (85) & & (86)
\end{array}$$

Thus with the acid (84) decarboxylation may well be promoted by the incipient transfer of proton through hydrogen bonding. If such is the case the pathway will proceed *via* the enol form (85) of the final end-product (86). Such enol $\longrightarrow$ keto conversions are generally fairly rapid and, unless special stabilizing features are present, tend to lie very far over in favour of the keto form. In the above case, however, the conversion is slow and the overall reaction can be followed by u.v. spectroscopy. The original absorption of (84) at 335nm ($\varepsilon = 38$) declines and is replaced by absorption at 240nm ($\varepsilon = 7500$)—interpreted as due to (85) by analogy with other species—which, in turn, slowly changes to absorption at 310nm due to (86). The presence of the enol (85) can also be demonstrated by its reaction with bromine

$$\underset{(85)}{\overset{\overset{\displaystyle OH}{|}}{Me_2C{=}C{-}CO_2H}} + Br_2 \longrightarrow \underset{(87)}{\overset{\overset{\displaystyle O}{\|}}{Me_2\,\underset{\underset{\displaystyle Br}{|}}{C}{-}C{-}CO_2H}} + H^{\oplus}Br^{\ominus}$$

to yield a brominated derivative (87) of the normal decarboxylation product (86), which is itself unaffected by bromine under these conditions as also is the starting material (84).

The generation of small ambient concentrations of carbanions (e.g. 88), by base-induced removal of proton from species in which the resultant carbanion is stabilized by delocalization, is well known:

$$HO^{\ominus}\curvearrowright H{-}\overset{\curvearrowleft}{\underset{\underset{\displaystyle H}{|}}{CH_2}}{-}C{=}O \rightleftharpoons H_2O + \left[\overset{\frown}{^{\ominus}CH_2}{-}\overset{\curvearrowleft}{\underset{\underset{\displaystyle H}{|}}{C}}{=}O \leftrightarrow CH_2{=}\underset{\underset{\displaystyle H}{|}}{C}{-}O^{\ominus}\right]$$

$$(88)$$

The formation of such carbanions, and their subsequent trapping by molecules of as yet unreacted starting material is indeed the basis of a number of preparatively useful reactions related to the aldol condensation:

$$\underset{\underset{\underset{\displaystyle (88)}{\displaystyle CH_2CHO}}{\overset{\displaystyle \ominus}{}}}{CH_3{-}\overset{\overset{\displaystyle |\delta+}{}}{\underset{}{C}}{\overset{\delta-}{=}}O} \rightleftharpoons \underset{\underset{\displaystyle CH_2CHO}{|}}{CH_3{-}\overset{\overset{\displaystyle H}{|}}{C}{-}O^{\ominus}} \overset{H_2O}{\rightleftharpoons} \underset{\underset{\displaystyle CH_2CHO}{|}}{CH_3{-}\overset{\overset{\displaystyle H}{|}}{C}{-}OH} \overset{-H_2O}{\longrightarrow} CH_3CH{=}CHCHO$$

Further support for such generation of carbanions is provided by the observed behaviour of species carrying a built-in trap in the form of a carbon atom, near enough to be attacked, that carries a good leaving group, i.e. one readily lost as an anion. Thus attack of methoxide ion on the α-bromoketone (89) is found to yield not the expected methyl ether (90) but the methyl ester (91):

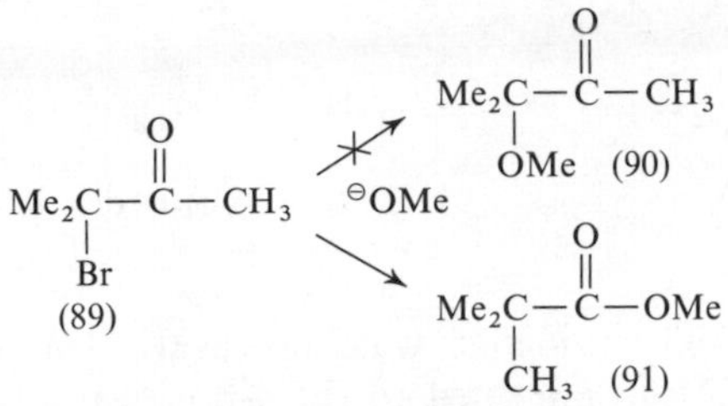

$$
\underset{\substack{\displaystyle | \\ Br \\ (89)}}{Me_2C}-\overset{\displaystyle O}{\overset{\displaystyle \|}{C}}-CH_3 \quad {}^{\ominus}OMe
$$

$$
\underset{\substack{\displaystyle | \\ OMe \quad (90)}}{Me_2C}-\overset{\displaystyle O}{\overset{\displaystyle \|}{C}}-CH_3
$$

$$
\underset{\substack{\displaystyle | \\ CH_3 \quad (91)}}{Me_2C}-\overset{\displaystyle O}{\overset{\displaystyle \|}{C}}-OMe
$$

This surprising result can be rationalized by a pathway such as

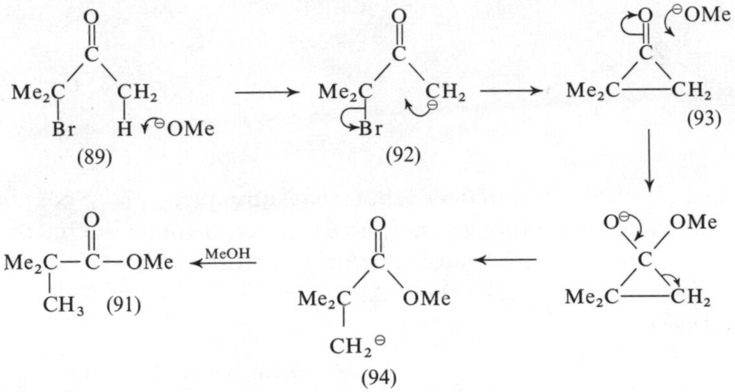

in which the first-formed carbanion (92) is trapped by the carbon atom
β-to it, with loss of bromide ion, to yield the cyclopropanone (93);
this, in turn, undergoes ready ring-opening by methoxide ion to yield
a second carbanion (94) which stabilizes itself by abstraction of a
proton from the solvent to yield the observed end-product (91): such
changes are generally referred to as Favorskii rearrangements. Further
detailed support for the pathway suggested above is provided subsequently (p. 208).

(f) In the Claisen rearrangement

Reference has already been made (p. 65) to the fact that if, in the
Claisen rearrangement of O-allylphenols, the *o*-positions are blocked
(95) the allyl group migrates to the vacant *p*-position (96):

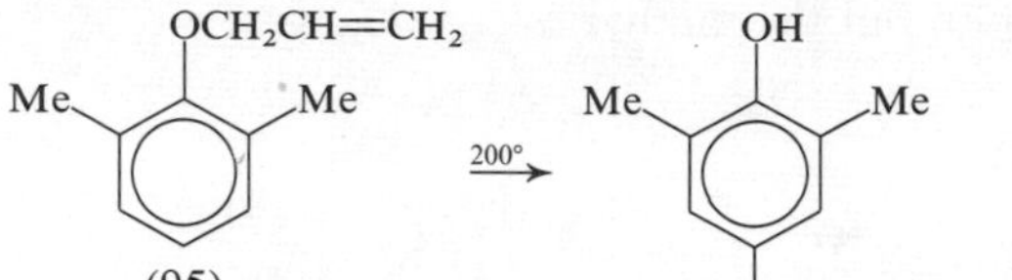

(95) $\xrightarrow{200°}$ (96)

Isotopic (^{14}C) labelling evidence was, however, then presented to show that the allyl group migrated to the *p*-position not directly but probably *via* an *o*-intermediate (97):

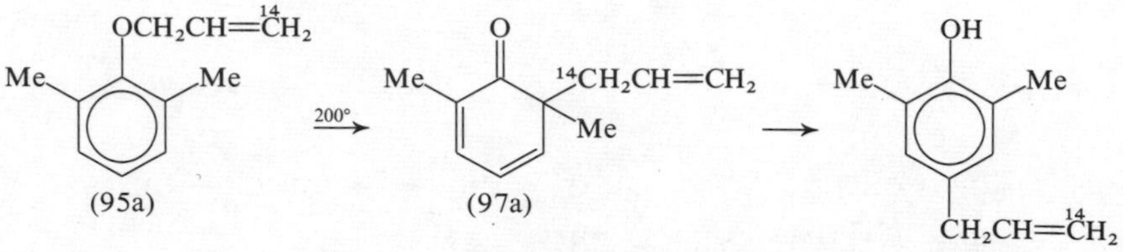

(95a) $\xrightarrow{200°}$ (97a) $\longrightarrow$ (96a)

This evidence was inferential only, but it has subsequently been possible, under suitable conditions, to trap the suggested dienone intermediate (97) by 1,4-addition of maleic anhydride across the carbon-carbon conjugated double bonds (Diels–Alder reaction), and to isolate the adduct (98):

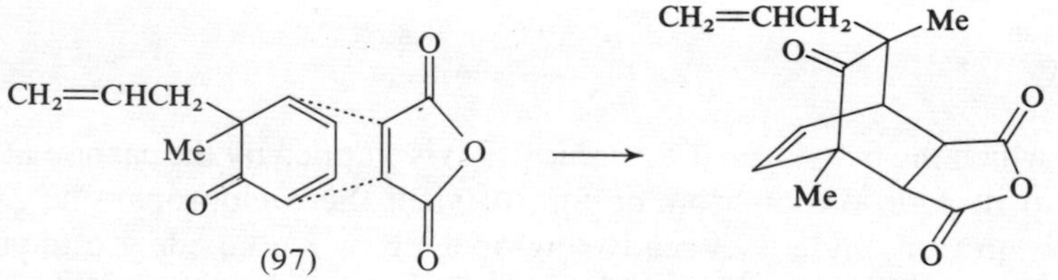

(97) $\longrightarrow$ (98)

On heating this adduct (98), the Diels–Alder reaction is reversed, but under the conditions necessary for this reversal the reformed (97) is converted into the normal *p*-allylphenol end-product (96).

(g) In ozonolysis

The addition of ozone to alkenes to form ozonides long resisted direct study because of the lability of the latter, which were often explosive and were thus usually decomposed *in situ* to yield the normal carbonyl end products:

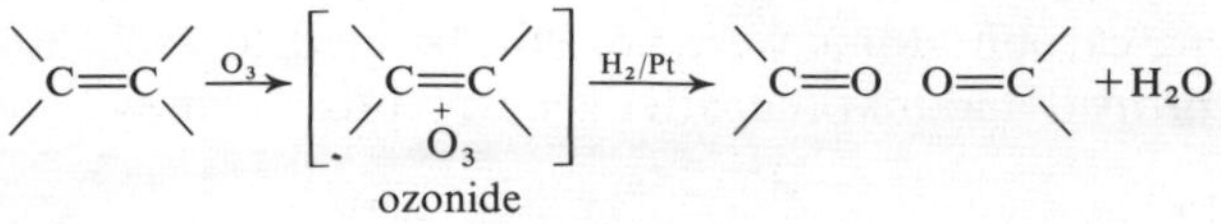

$$\overset{\diagdown}{\diagup}C = C\overset{\diagup}{\diagdown} \xrightarrow{O_3} \left[\overset{\diagdown}{\diagup}\underset{\overset{+}{O_3}}{C} = C\overset{\diagup}{\diagdown} \right] \xrightarrow{H_2/Pt} \overset{\diagdown}{\diagup}C = O \quad O = C\overset{\diagup}{\diagdown} + H_2O$$

ozonide

Initial attack of ozone on the double bond could be envisaged as a 1,3-bipolar addition initiated by the electrophilic end of the ozone dipole (99),

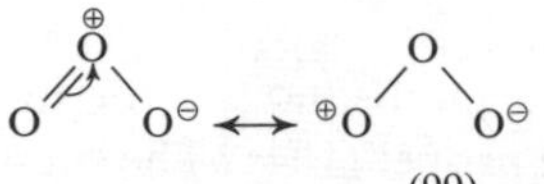

(99)

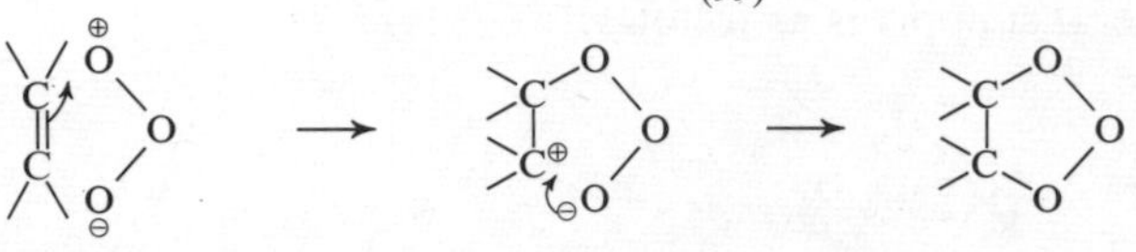

(99)   (100)

and this is supported by the fact that the addition of ozone is promoted by Lewis acid catalysts such as BF_3. It seems unlikely that the ozonide proper has structure (100), however, as this would not be expected to undergo the very ready decomposition to two molecules of carbonyl compound on hydrolysis or hydrogenation, under extremely mild conditions, as the true ozonide is observed to do. Further, n.m.r. spectra have shown in a number of cases that the first-formed adduct (*molozonide*) decomposes rapidly to yield the ozonide proper. Support for this first-formed adduct having structure (100) is provided by the actual isolation, under very mild conditions, of the species (102) from the action of ozone on *trans*-2,2,5,5-tetramethylhex-3-ene (101), the observation of only a single C—H (as opposed to CH_3) signal in its n.m.r. spectrum—i.e. the two protons of this type are therefore, symmetrically disposed—and the demonstration that the adduct on reduction yields the 1,2-diol (103)—i.e. the original carbon–carbon double bond is not yet wholly cleaved in the adduct:

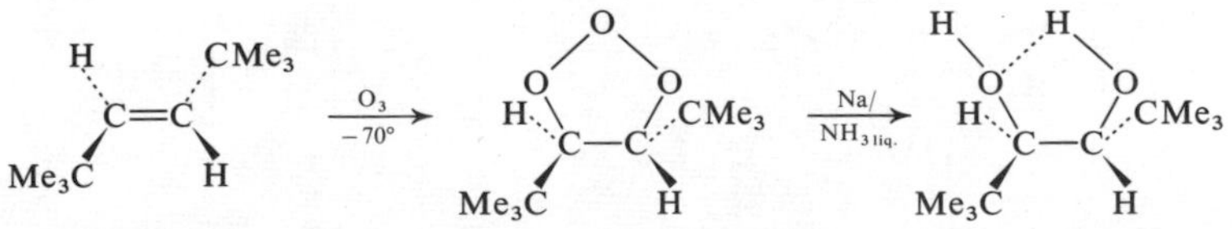

(101)   (102)   (103)

This last observation clearly differentiates the initial adduct (102) from the ozonide proper for the latter, on reduction, yields two molecules of carbonyl compound. It had indeed long been thought that the two alkene carbon atoms were not still bonded to each other in the ozonide proper, and that the latter is best represented by (104):

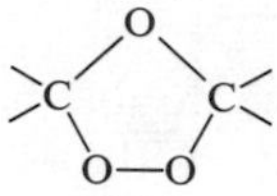

(104)

The question therefore remains, how is the initial adduct (100) transformed into the ozonide proper (104)? The suggestion commanding the most general support is as follows:

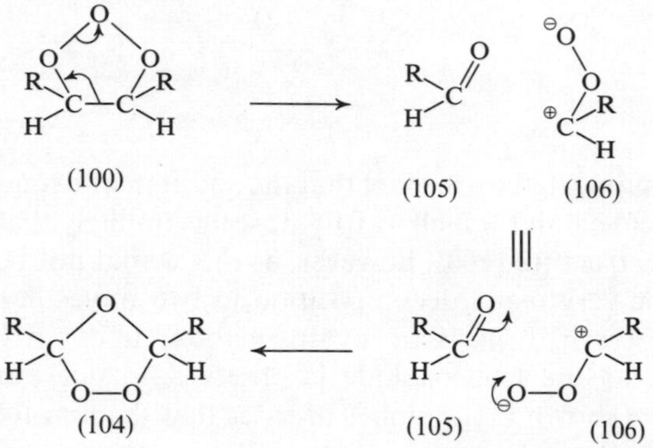

(100)   (105)   (106)

|||

(104)   (105)   (106)

This involves initial fission of (100) to yield the peroxyzwitterion (106) and a carbonyl compound, in this case an aldehyde, (105). Recombination of these fragments, initiated by nucleophilic attack of the terminal oxygen atom of (106) on the carbonyl carbon atom of (105), then yields the true ozonide.

Support for such a reaction pathway is provided by the fact that products obtainable from peroxyzwitterions (106), i.e. cyclic peroxide dimers (107), and polymers (108),

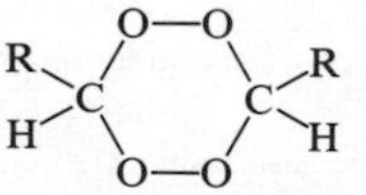

(107) 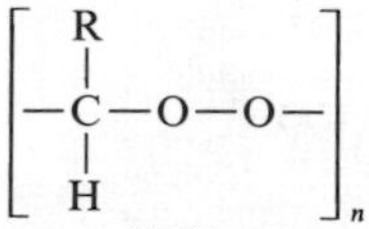   (108)

are indeed obtained as by-products in the normal ozonolysis of alkenes. Further, such a conversion should, in theory, lend itself well to confirmation by trapping, for carrying out the ozonization in the presence of a 'foreign' aldehyde, R'CHO, should—provided the latter does not differ too markedly from RCHO—lead to the formation of a mixed ozonide (109), through interaction of R'CHO with the peroxyzwitterion (106),

$$R' \underset{H}{\overset{O}{\diagup}} \underset{O-O}{\overset{|}{\diagdown}} \underset{H}{\overset{R}{\diagup}}$$

(109)

in addition to the normal, symmetrical ozonide (104) derived from the 'internal' aldehyde (105). Such mixed ozonides may indeed be detected and, coupled with the peroxyzwitterion by-products mentioned above, constitute considerable evidence in favour of a conversion involving (105) and (106).

It should also be theoretically possible, during ozonolysis of an unsymmetrical alkene, to detect 'internal trapping' by identification of two symmetrical ozonides in addition to the expected unsymmetrical one; this has been demonstrated with 2-pentene (110):

MeCH=CHCH$_2$Me

(110)

| $\downarrow$ O$_3$

$$\underset{(111)}{\overset{O}{MeCH-CHCH_2Me}}$$

$$\longrightarrow \left[\underset{(112)}{MeCHO} + \underset{(113)}{^{\ominus}O-O-\overset{\oplus}{C}HCH_2Me} \atop \underset{(114)}{Me\overset{\oplus}{C}H-O-O^{\ominus}} + \underset{(115)}{OHCCH_2Me} \right]$$

$$\underset{(117)}{\overset{O}{MeCH \quad CHMe}} \quad \text{From: } 112 + 114 \quad \nearrow$$

$$\underset{(118)}{\overset{O}{MeCH_2CH \quad CHCH_2Me}} \quad \text{From: } 115 + 113$$

$$\longrightarrow \underset{(116)}{\overset{O}{MeCH \quad CHCH_2Me}} \quad \begin{array}{l} \text{From: } 112 + 113 \\ 115 + 114 \end{array}$$

The ozonides of 2-butene (117) and 3-hexene (118) may indeed be isolated in addition to the one expected from 2-pentene (116) and, as might be expected, the proportion of the 'symmetrical' ozonides obtained decreases with dilution of the original solution of alkene: the aldehyde/zwitterion fragments (112 + 113) from the decomposition of one molecule of molozonide (111) are less likely to collide with the (alternative) fragments (114 + 115) from another molecule as dilution increases and are, therefore, increasingly likely to recombine with each other. Very little cross-ozonide formation is indeed found to occur at the dilutions generally employed for preparative or diagnostic ozonolysis. Similar cross-ozonide formation may also be observed in the simultaneous ozonolysis of two symmetrical alkenes present together in the same solution, though here of course it is the 'unsymmetrical' ozonide that we are looking for, e.g. ozonolysis of a mixture of 3-hexene and 4-octene is found to yield some 3-heptene ozonide. These 'crossover' experiments (*cf.* p. 59) have only been made possible by the use of gas-liquid partition chromotography to separate successfully the very closely similar ozonides.

It should be said that though this reaction pathway for ozonolysis is satisfactory as far as it goes, it is not able to explain the detailed stereochemistry of the reaction, i.e. the relationship between the stereochemistry (*cis/trans*) of the original alkene and that (*cis/trans* of the substituents on the—ultimately—carbonyl carbons) of the product ozonides; it will, therefore, need some modification as more refined data become available.

(h) In 'crossover' experiments

The 'crossover' experiments referred to above in ozonolysis, and those already mentioned in connection with the establishment of the intramolecularity of the benzidine rearrangement (p. 59), serve by internal, crossover, trapping to indicate the formation of an intermediate, and may (as above) provide some information about that intermediate's nature. They have been found particularly valuable in establishing whether rearrangement reactions are *inter*- or *intra*-molecular (*cf.* p. 58), i.e. whether a migrating group actually becomes free during the course of the reaction (*inter*-) or whether it does not (*intra*-). The benzidine rearrangement was thus shown to be wholly intramolecular, but this is not necessarily true of all such reactions. Thus although the rearrangement of phenolic esters (119) to hydroxyphenones (120, 121), the Fries rearrangement,

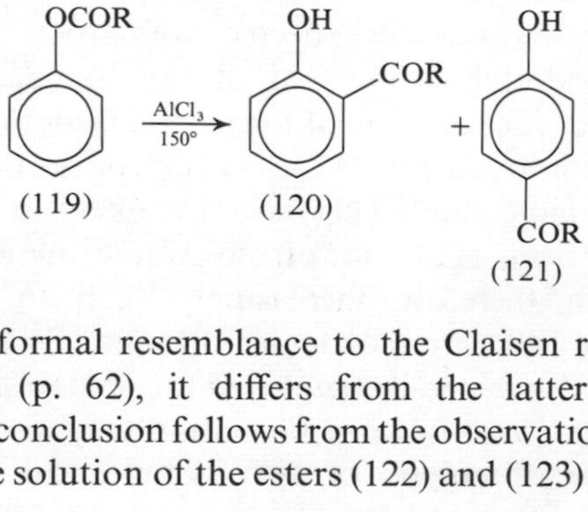

bears a certain formal resemblance to the Claisen rearrangement of O-allyl phenols (p. 62), it differs from the latter in being intermolecular. This conclusion follows from the observation that rearrangement in the same solution of the esters (122) and (123) leads not only to the expected products (124) and (125), but also to the 'crossed products' (126) and (127):

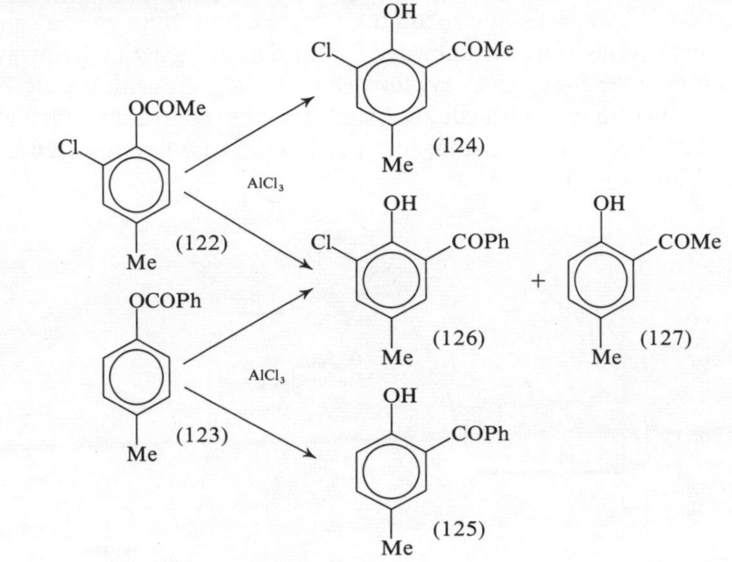

The nature of the necessary catalyst—$AlCl_3$, a Lewis acid—coupled with the proven inter-molecularity suggests that the reaction probably involves initial decomposition of the ester by $AlCl_3$ (known from other reactions), followed by Friedel–Crafts acylation of the activated benzene nucleus (electrophilic substitution) by an acyl carbonium ion (128), an extremely powerful electrophile:

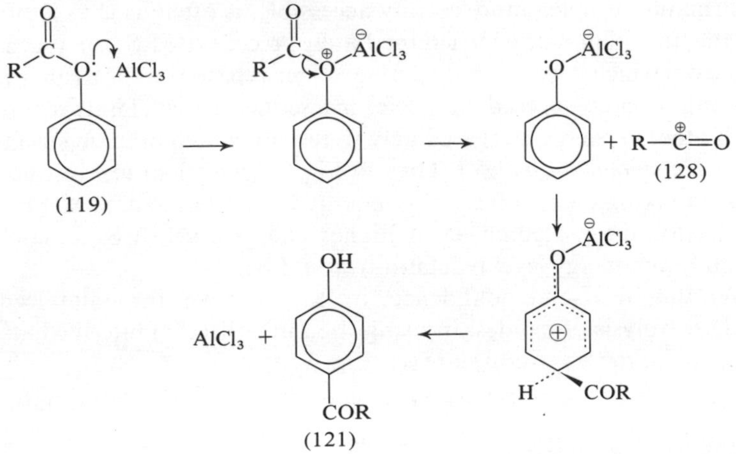

INTERMEDIATES AS MODELS FOR TRANSITION STATES

As will have been seen from the foregoing examples, the recognition of particular species as intermediates in a reaction often goes a long way to provide an overall framework for the detailed reaction pathway. In defining a reaction pathway, however, it is the transition state or states, rather than intermediates, that are the controlling factors, particularly the transition state for the slowest, rate-limiting step i.e. $T.S_1.$, below (Fig. 3.2):

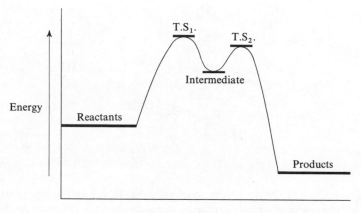

Fig. 3.2

Ideally, we should most like to have information about such transition states, but this is often very difficult to obtain directly or is even not obtainable at all. In such circumstances an intermediate, about which information is much more readily accessible, is often used as a model for the transition state which immediately precedes it ($T.S_1.$, in Fig. 3.2), and from which it was directly derived. This is based on the assumption that immediately succeeding species in a sequence that closely resemble each other in energy level are likely to resemble each other in structure also (*Hammond's principle*). Thus in Fig. 3.2 we might feel justified at least in taking the relatively high energy level intermediate as a better model for the preceding, even higher energy level, $T.S_1.$, than the much lower energy level reactant(s) would be.

We thus feel some confidence, in, for example, the unimolecular (S_N1) solvolysis of halides, in using the ion pair (129) intermediate as a model for the transition state (*cf.* Fig. 3.2):

$$Me_3C-Br \xrightarrow[slow]{} [T.S_1.] \longrightarrow Me_3C^{\oplus}Br^{\ominus} \xrightarrow[fast]{H_2O:} Me_3C-OH + H^{\oplus}Br^{\ominus}$$
$$(via\ T.S_2.)$$
$$(129)$$

We could then forecast, under parallel conditions, the reactivity sequence

$$Me_3C—Br < PhCMe_2—Br < Ph_2CMe—Br < Ph_3C—Br$$

on the basis of the relative stabilization, by delocalization, of the carbonium ion moiety of the ion pair intermediates in each case:

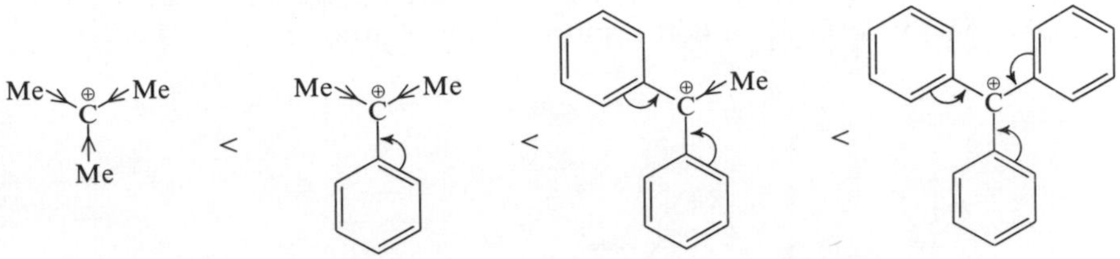

The assumed parallel, and progressive, stabilization of the corresponding transition states (i.e. lowering of their energy levels) would be expected to lead to their readier attainment, and so to a progressive increase in the rate of the reaction as the series is traversed.

Another example is the use of σ complex intermediates as models for the transition states in aromatic electrophilic substitution, e.g. in the nitration of anisole (methoxybenzene) in order to forecast the most likely products. There is, of course, the possibility of attack by $^{\oplus}NO_2$ (*cf.* p. 90) at three different positions, leading to *o*-, *m*-, and *p*-products, and hence three σ-complexes to consider:

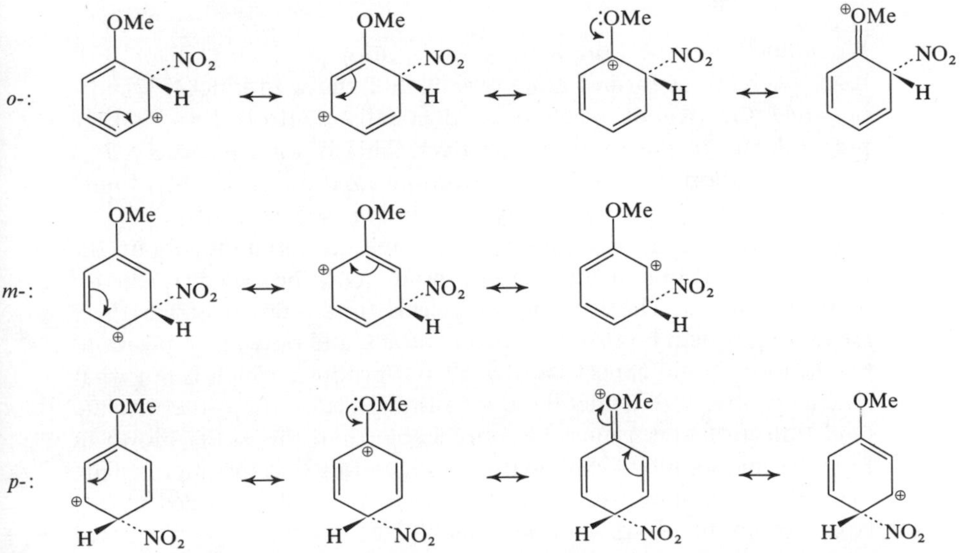

In considering the relative stabilization, by delocalization, of these intermediates an additional recourse is open to the species resulting from *o*- and *p*- (but not from *m*-) attack involving an electron pair on the oxygen atom of the OMe group, thereby furthering the delocalization in, and hence relative stabilization of, these two species. The intermediates, and by inference the immediately preceding transition states, resulting from *o*- and *p*-attack are thus more stable, i.e. at a lower energy level, than that resulting from *m*-attack:

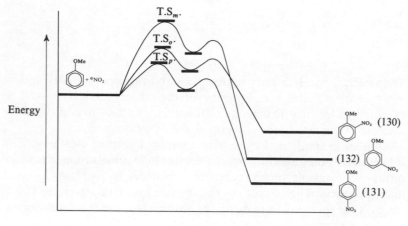

Fig. 3.3

We should thus expect the *o*- and *p*-substitution products (130 and 131, respectively) to be formed more rapidly than the *m*-product (132) and, provided the overall reaction is kinetically controlled (p. 35), to preponderate in the total end product. This is what is observed: a typical nitration of anisole with nitrating mixture at 45° was found to lead to *o*- 31 per cent, *p*- 67 per cent, and *m*- ≈ 2 per cent.

This simple consideration does not tell us anything about the relative proportions of the *o*- and *p*-products to be expected (the *o-/p*-ratio), however. The distribution is clearly not a statistical one (there are two equivalent *o*-positions open to attack and only one *p*-position) for then we should expect more *o*- than *p*-product, which is not what is actually observed. Under these conditions at least, the *p*-intermediate (and transition state) must be more stable than the *o*- (as shown in Fig. 3.3), and we might explain this by suggesting that the intermediate (and transition state) for *o*-attack is more sterically crowded by the large groups involved, and so destablilized (*cf.* p. 116), compared with that for *p*-attack. The *o*-/*p*-ratio is greatly affected by the conditions of the reaction, however, and no simple treatment leads to any very satisfactory predictions of its magnitude.

FURTHER READING

BETHELL, D. 'Structure and Mechanism in Carbene Chemistry', *Adv. Phys. Org. Chem.* (Academic Press), 1969, **7**, 153.

BETHELL, D. and GOLD, V. *Carbonium Ions: An Introduction* (Academic Press, 1967).

CRAM, D. J. *Fundamentals of Carbanion Chemistry* (Academic Press, 1965).

CRAM, D. J. 'Carbanions', *Surv. Progr. Chem.* (Academic Press), 1968, **4**, 45.

DYKE, S. F., FLOYD, A. J., SAINSBURY, M. and THEOBALD, R. S. *Organic Spectroscopy: An Introduction* (Penguin, 1971).

GILCHRIST, T. L. and REES, C. W. *Carbenes, Nitrenes and Arynes* (Nelson, 1969).

HUISGEN, R. 'Kinetic Evidence for Reactive Intermediates', *Angew. Chem. Int. Ed.,* 1970, **9**, 751.

JENCKS, W. P. 'Mechanism and Catalysis of Simple Carbonyl Group Reactions', *Progr. Phys. Org. Chem.* (Interscience), 1964, **2**, 63.

LWOWSKI, W. (Ed.) *Nitrenes* (Interscience, 1970).

MURRAY, R. W. 'The Mechanism of Ozonolysis', *Acc. Chem. Res.,* 1968, **1**, 313.

NORMAN, R. O. C. and GILBERT, B. C. 'Electron-spin Resonance Studies of Short-lived Organic Radicals', *Adv. Phys. Org. Chem.* (Academic Press), 1967, **5**, 53.

OLAH, G. A. 'Mechanism of Electrophilic Aromatic Substitutions', *Acc. Chem. Res.,* 1971, **4**, 240.

OLAH, G. A. and PITTMAN, C. U. 'Spectroscopic Observations of Alkylcarbonium Ions in Strong Acid Solution', *Adv. Phys. Org. Chem.* (Academic Press), 1966, **4**, 305.

OLAH, G. A. and SCHLEYER, P. VON R. (Eds.) *Carbonium Ions, Vols. I, II* (Interscience, 1968, 1970).

TEDDER, J. M. 'The Interaction of Free Radicals with Saturated Aliphatic Compounds', *Quart. Rev.,* 1960, **14**, 336.

WALLING, C. (Ed.) *Free Radicals in Solution* (Wiley, 1957).

WILLIAMS, D. H. and FLEMING, I. *Spectroscopic Methods in Organic Chemistry* (McGraw Hill, 1966).

WINSTEIN, S., APPEL, B., BAKER, R. and DIAZ, A. 'Ion Pairs in Solvolysis and Exchange', *Organic Reaction Mechanisms,* Special Pub. No. 19 (Chem. Soc., London), 1965, pp. 109–130.

4
Stereochemical criteria

Probably the most familiar stereochemical influence on reaction pathways is the classical 'steric hindrance' in which the sheer bulk of substituents impedes the approach of a reagent to the reacting position in a molecule, and results in the formation of a highly crowded—and hence high-energy—transition state such that the reaction proceeds only very slowly if at all.

114

BULK EFFECTS

This effect is seen clearly in the relative rates of bimolecular (S_N2) reaction of the following halides with ethoxide ion in ethanol:

CH_3CH_2—Br	$MeCH_2CH_2$—Br	Me_2CHCH_2—Br	Me_3CCH_2—Br
(1)	(2)	(3)	(4)
1	$2{\cdot}8 \times 10^{-1}$	$3{\cdot}0 \times 10^{-2}$	$4{\cdot}2 \times 10^{-6}$

Each is a primary halide, kinetic measurements show that there is no mechanistic change as the series is traversed, and any electronic effect arising from the successive introduction of methyl groups on the β-carbon atom is likely to be small; the interesting point is that while the rate difference between the first and second, and between the second and third, halides is less than ten-fold in each case, that between the third and last halides is no less than ten thousand-fold! The reason for this sudden, enormous drop in rate—rather than a smooth, progressive decrease across the series—becomes apparent when we consider the approach of ethoxide ion to the α-carbon atom along the line of the bond joining that carbon atom to bromine, but from the side opposite to the latter as the S_N2 mechanism requires (*cf.* p. 126). Thus with the halide (3).

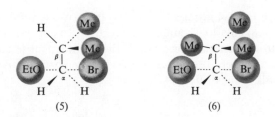

the S_N2 transition state will certainly be somewhat crowded, but rotation about the C_α—C_β bond will allow of one conformation (5), in which the approach of $EtO^\ominus$ is impeded only by H rather than by Me, and in which crowding in the forming transition state will be minimal (there will of course be two such conformations possible with (2) and three with (1)). By contrast *no* such conformation is available to (4), the approach of $EtO^\ominus$ will be difficult in the extreme, and the transition state will be of much higher energy (6): the $\Delta G^{\pm}$ for the reaction will thus be much higher and the reaction itself correspondingly slower. A similar resistance to bimolecular displacement in a

primary (di)halide with three β-substituents has already been encountered in the isolation from strongly basic solution of the species
(7) from the Reimer-Tiemann reaction of *p*-cresol (p. 95):

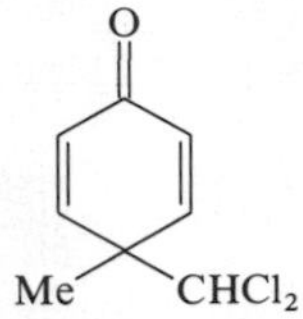

(7)

The operation of straightforward bulk effects is also seen in the
relatively low reactivity of the blue radical (9) obtained by the ferri-
cyanide oxidation of the phenol (8):

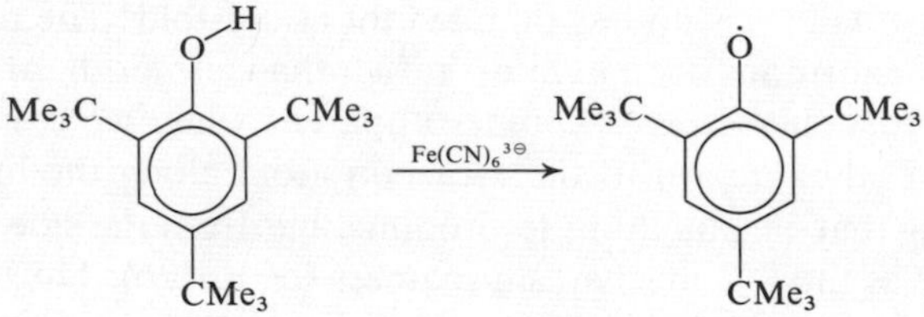

(8)   (9)

The very large *t*-butyl groups in the two *o*-positions inhibit the close
approach of reagents to Ȯ and preserve the radical, especially from
the usual fate of dimerization. Much the same effect is seen in, for
example, the nitration of a series of alkylbenzenes that have progressively larger alkyl substituents (10–13),

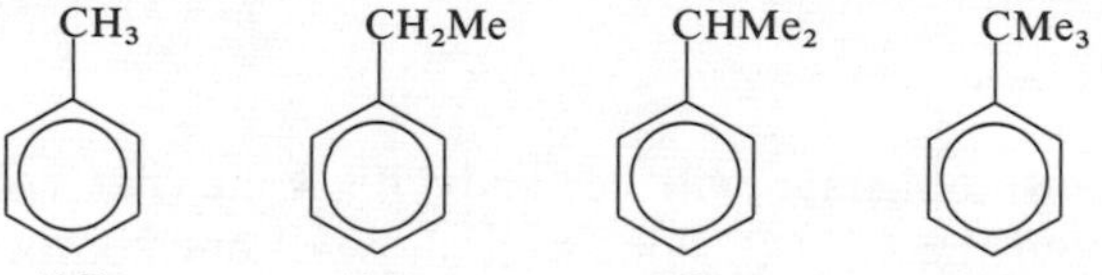

(10)   (11)   (12)   (13)

where the proportion of *o*-nitro product drops from 57 per cent with
toluene (10) to 12 per cent with *t*-butylbenzene (13), while the proportion of *p*-nitro product rises from 40 per cent (with 10) to 80 per cent
(with 13). While in this case electronic effects probably differ little
along the series, and the changed *o*-/*p*-ratio stems almost wholly from
steric influences in the transition state involved in attack on the *o*-

position, the same clearly cannot be true for nitration of the halogeno-benzenes (14–17):

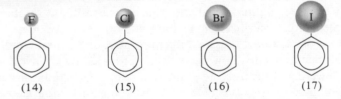

| (14) | (15) | (16) | (17) |

Here the proportion of *o*-nitro product *increases* from 12 per cent (with 14) to 38 per cent (with 17) despite an increase in the size of the substituent already present: clearly electronic effects must also be involved.

A further well-documented example is the failure of mesitoic acid (18) to esterify under normal conditions, e.g. excess of alcohol plus an acid catalyst, no matter how long the heating under reflux is continued for:

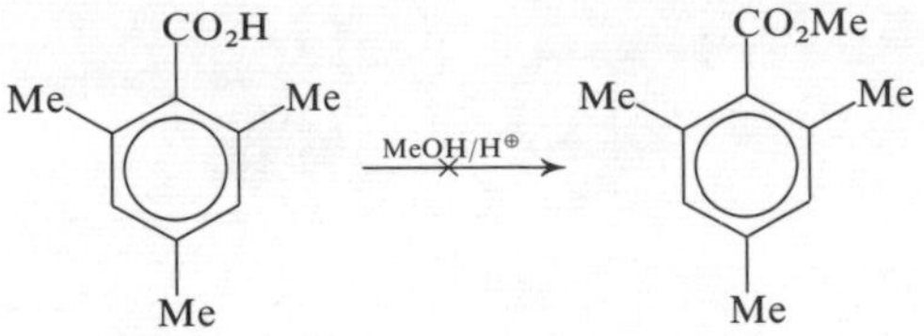

Initial protonation could take place, for this is of low steric demand, but the planar, protonated carboxyl group would be forced by the two bulky *o*-methyl substituents to take up the conformation (19),

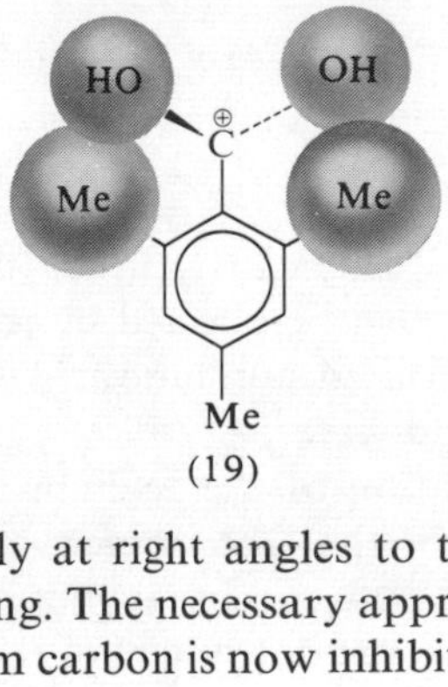

in which it is essentially at right angles to the plane of the carbon atoms in the benzene ring. The necessary approach of the nucleophilic MeÖH to the carbonium carbon is now inhibited from all directions— from below by the benzene ring, from above and from two sides by

OH groups, and from the last two sides by the *o*-methyl groups— steric hindrance is thus total!

It is, however, possible to esterify the acid (18) by dissolving it in concentrated sulphuric acid and pouring the resultant solution into cold methanol; esterification is then substantially complete. Steric inhibition of reaction is thus no longer operative under these conditions, and the clue to what is now occurring is provided by an investigation of the lowering of the freezing point of concentrated sulphuric acid (*cf.* p. 90) induced by dissolving (18) in it. An almost four-fold depression is observed and this may be interpreted in terms of the following equilibrium:

$$\underset{(18)}{Ar\overset{\underset{\displaystyle |}{OH}}{C}=O} + 2H_2SO_4 \rightleftharpoons \underset{(20)}{Ar\overset{\oplus}{C}=O} + H_3O^\oplus + 2HSO_4^\ominus$$

The acylium cation (20) will have the following structure

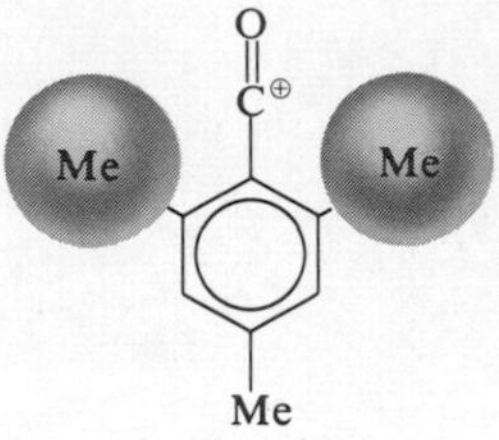

(20)

in which attack by nucleophilic MeÖH on the carbonium ion carbon is now substantially unimpeded from the back or front of the molecule; esterification thus takes place:

$$\underset{(20)}{Ar\overset{\oplus}{C}=O} + MeOH \rightleftharpoons \underset{(21)}{Ar\overset{\underset{\displaystyle |}{OMe}}{C}=O} + H^\oplus$$

Hydrolysis of mesitoate esters (21), which does not proceed under normal conditions, can also be effected by reversal of the above, i.e. dissolving (21) in concentrated sulphuric acid to form (20), and pouring the resulting solution into water.

The reason for formation of the acylium ion (20) rather than the protonated species (19) is that the former—or more particularly the transition state leading to it—can be stabilized by delocalization of its positive charge over the π orbital system of the benzene ring while the latter cannot:

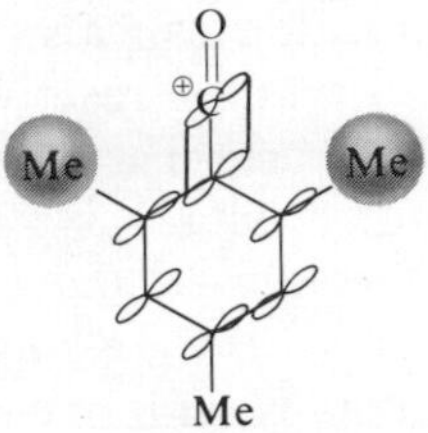

(20)

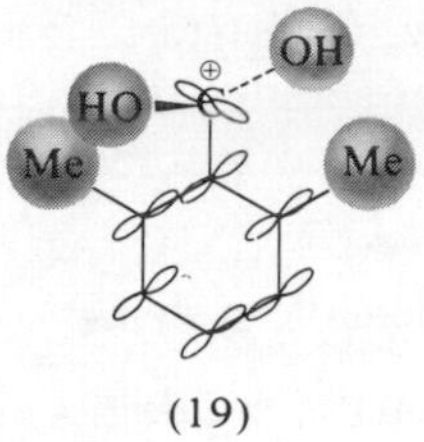

(19)

This inhibition of delocalization occurs because the *p*-orbital on the carbonium ion carbon is prevented—by steric interaction of the OH and Me groups—from taking up a position that is anywhere near parallel to the *p* orbital on the adjacent ring carbon atom. Overlap between them thus cannot take place, and the result is steric inhibition of delocalization.

It is significant in this respect that dissolving the unsubstituted benzoic acid (22) in concentrated sulphuric acid is found to result in only a two-fold depression of freezing point:

$$\underset{(22)}{C_6H_5\overset{\overset{\displaystyle OH}{|}}{C}{=}O} + H_2SO_4 \rightleftarrows \underset{(23)}{C_6H_5\overset{\overset{\displaystyle OH}{|}}{\underset{\oplus}{C}}{-}OH} + HSO_4^{\ominus}$$

The simple protonated species (23) is formed

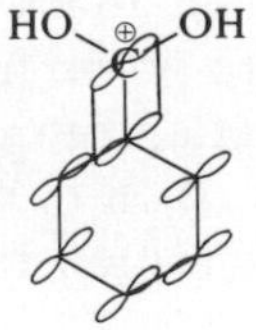

(23)

because *p/p* orbital overlap is no longer sterically inhibited, and delocalization of the positive charge, with consequent stabilization, can now take place: there is thus no need, as there is with (19), for initial bond-breaking to yield an acylium cation, $C_6H_5C^{\oplus}{=}O$, before stabilization through *p/p* orbital overlap can take place.

STERIC INHIBITION OF DELOCALIZATION

The example of (19) above involves a species that cannot become stabilized in this form because the expected delocalization is inhibited sterically. Such steric inhibition of delocalization can also operate to inhibit stabilization of an intermediate, so that the reactivity of a position or group is reduced apparently by steric action at a distance!

(a) In aromatic systems

The aromatic halide (24) is found to react readily with nucleophiles, e.g. piperidine, undergoing displacement of the bromine atom (*cf.* p. 80):

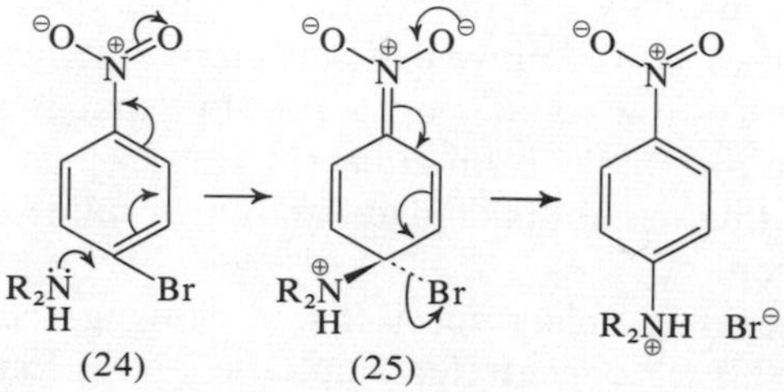

The effect of the nitro-group is to stabilize the intermediate (25), or more particularly the transition state that precedes it, by delocalization of the negative charge developed in forming a bond (donation of an electron pair) from the attacking nucleophile to the ring carbon atom. By contrast, the halide (26), which should also be activated by NO_2, reacts at only one fortieth the rate of (25) under parallel conditions; despite the fact that the substituent methyl groups are too far away to exert any steric effect directly on attack by the nucleophile. They act, in fact, by preventing the nitro-group from becoming co-planar with the benzene ring, and thus overlap between the *p* orbitals of the nitro-nitrogen atom and the ring carbon to which it is attached is greatly reduced, as therefore is stabilization in the intermediate (27) compared with (25):

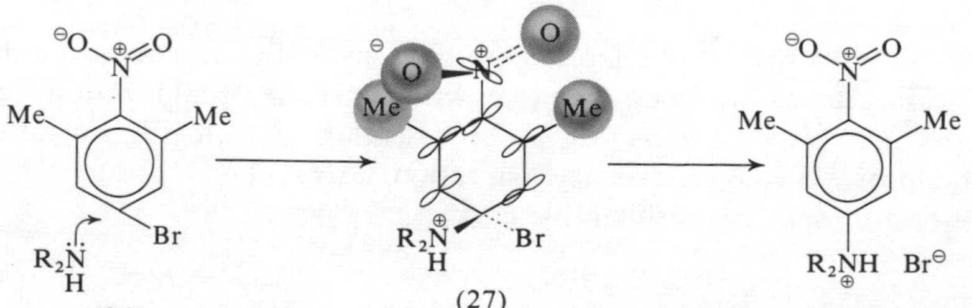

A similar but more pronounced effect is seen with the pair of amines (28) and (29),

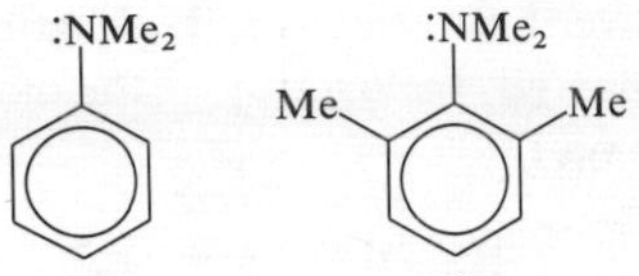

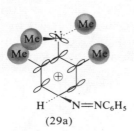

(28)   (29)

in that while the first of them (28) will couple readily with benzene diazonium chloride,

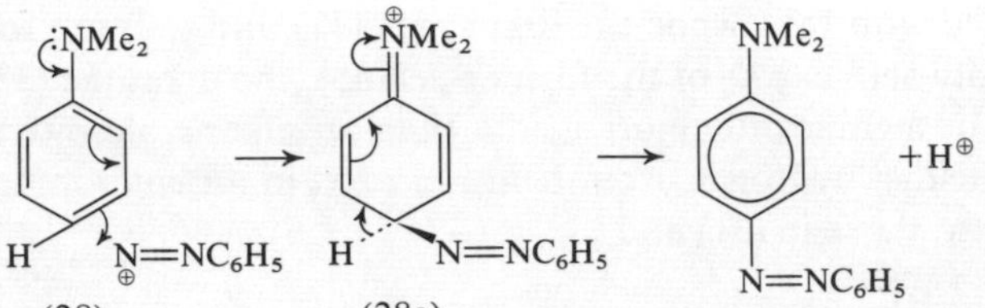

(28)   (28a)

due to stabilization of the intermediate (28a) through the agency of the electron pair on the NMe_2 group, the latter (29) does not couple at all under parallel conditions. Reaction fails to occur because delocalization, with consequent stabilization, is so inhibited (inability to achieve coplanarity) in the potential intermediate (29a)—and, more importantly, in the transition state that would precede it—that its formation is too energy-demanding; the reaction does not therefore occur under normal conditions:

(29a)

(b) In carbanions

A similar effect arising in a slightly different way, and this time influencing an equilibrium, is seen on comparing the acidities of a series of β-diketones:

(30)

Such compounds owe their relatively high acidity to the stabilization by delocalization, through the agency of the carbonyl groups, that can occur in the carbanion (30):

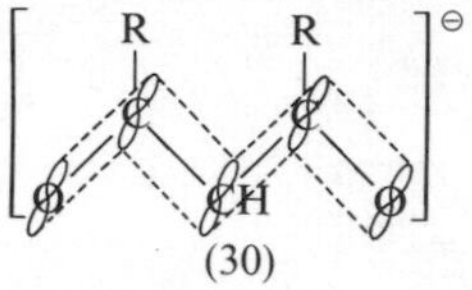

(30)

In line with this, while the diketones (31) and (32) are found to be markedly acidic (pK_a of the former is 8·65), the diketone (33) is found not to be significantly more acidic than an alkane, despite the potentially 'acidic' hydrogen atom being located in an environment closely similar to those in (31) and (32):

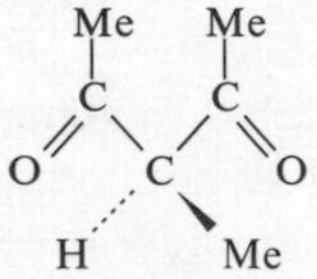

(31) 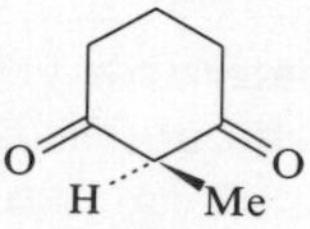   (32) 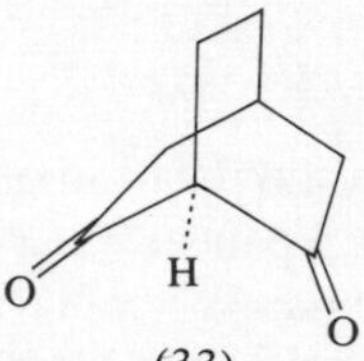   (33)

This unexpected absence of acidic character arises from the fact that stabilization by delocalization cannot occur in the carbanion (34) that would be obtained from (33): it just does not form therefore. The rigid, cage-like carbon framework prevents the carbanion carbon from 'collapsing' to a planar state, and its electron pair thus remains in an sp^3 orbital more or less at right angles to the p orbitals on the two carbonyl carbon atoms—sp^3/p overlap thus cannot take place and, consequently, no stabilization by delocalization will occur:

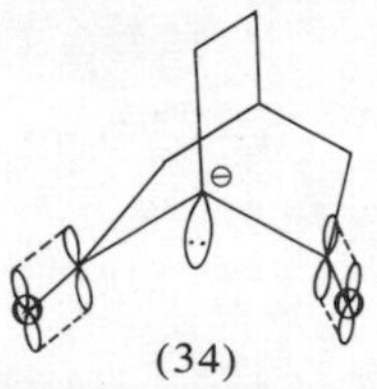

(34)

Even if the carbon framework could be distorted sufficiently to effect coplanarity of the bonds about the carbanion carbon atom, the resultant p orbital that would then contain the electron pair would still be essentially at right angles to the p orbitals on the adjacent carbonyl carbon atoms: delocalization could thus not take place even then. With or without deformation of the carbon framework, we are again seeing steric inhibition of delocalization.

It should be mentioned that carbanions, unlike carbonium ions (see below), appear to have no particular need, on energetic or other grounds, to assume a planar conformation but, like tertiary amines (35) with which they are iso-electronic, are pyramidal: one enantiomer being very readily and rapidly converted into the other by passing, momentarily, through a planar state:

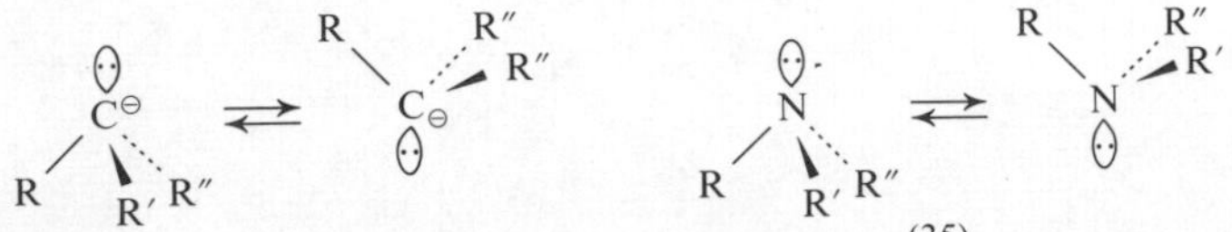

(35)

(c) In carbonium ions

An essentially analogous inhibition of delocalization, and hence of stabilization, may be observed in some potential carbonium ions. Thus triphenylmethyl bromide (36) undergoes solvolysis by a unimolecular pathway (S_N1) very, very much more rapidly ($\approx 10^{23}$ times) than 1-bromotriptycene (37), despite the fact that the local environment of the C—Br linkage is the same in both cases, i.e. the carbon atom is attached to three benzene rings:

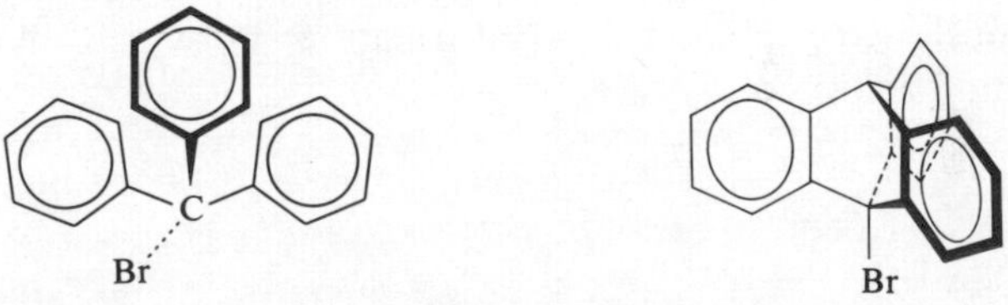

(36)   (37)

This enormous rate difference occurs because the carbonium ion intermediate (38)—and more particularly the transition state preceding it—that is derived from (36) can stabilize itself by delocalization through interaction with the π electron systems of the benzene rings:

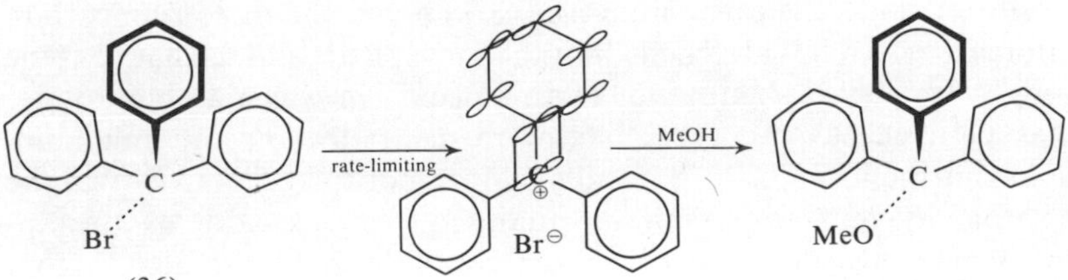

(36)   (38)

By contrast, the carbonium ion (39) that would be formed from 1-bromotriptycene (37) is unable (like the carbanion (34) above) to stabilize itself by delocalization, as overlap between the π orbital systems of the aromatic rings and the sp^3 orbital on the carbonium ion carbon atom cannot take place—they are virtually at right angles to each other; (39) thus just does not form:

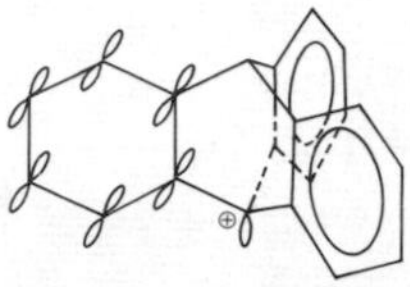

(39)

Here too, there would be no advantage if the cage-like structure could be distorted sufficiently to allow the carbonium ion carbon atom to become sp^2 hybridized for its p orbital would still be virtually at right angles to those on the aromatic nuclei.

Two minor points need to be made in connection with (36 $\longrightarrow$ 38) and (37). Firstly, the delocalization stabilization of (38) is not, in practice, quite as great as, in theory, it could be; due to steric interaction between o-H atoms on adjacent nuclei, this species cannot take up the conformation in which all nineteen carbon atoms are simultaneously co-planar—the conformation in which maximum delocalization could occur:

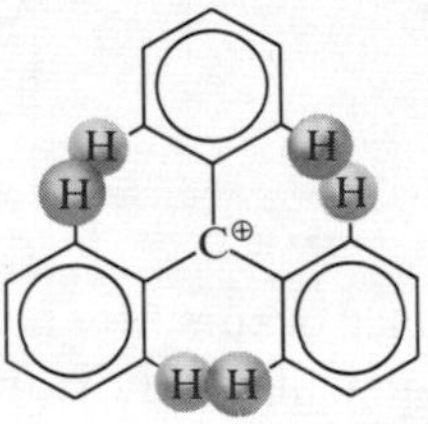

(38a)

This carbonium ion has actually been shown to have a propellor-like conformation—the carbonium ion carbon and the three ring carbon atoms to which it is attached all in the same plane, and the flat benzene rings then angled slightly out of this plane—in which a compromise has been effected between maximum delocalization and minimum o-H interaction (38a). Secondly, 1-bromotriptycene (37) will be profoundly resistant to bimolecular (S_N2) attack also, as the approaching nucleophile is prevented by the cage-like structure from reaching the back of the carbon atom to which bromine is bonded.

A slightly less marked, but still extremely powerful, effect is seen to operate in the solvolysis (in 80 per cent ethanol at 25°) of the tertiary halides—2-bromo-2-methylpropane (40) and the bromide (41), whose relative rates of reaction are recorded below:

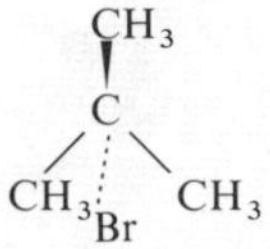

(40)
1

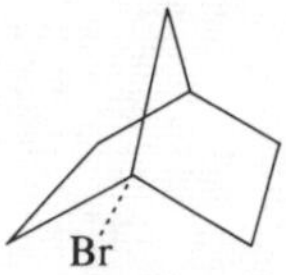

(41)
$\approx 10^{-13}$

Here again the local environment of the C—Br linkage is in each case formally analagous, but this time there is no question of p orbitals on adjacent aromatic systems and of possible overlap with them. The great difference in rate behaviour between the two bromides is believed to result from the fact that the carbonium ion (42) derived from (40) can collapse to the planar (sp^2) state, whereas the ion (43) that would be obtained from (41) cannot:

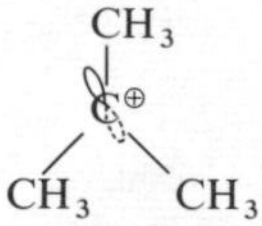

(42) 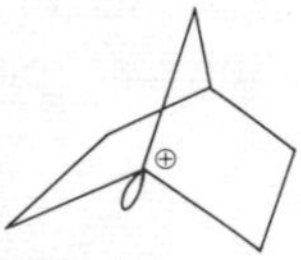 (43) 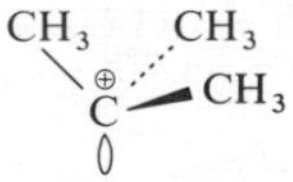 (42a)

It has been calculated that $Me_3C^{\oplus}$ is more stable in the planar (42), compared with the pyramidal (42a), state by $\approx 84kJ/mole$; if the same energy difference applied to the alternative transition states preceding these carbonium ion intermediates then the pathway involving the planar intermediate would clearly be very much the faster: the above energy difference would suggest relative rates of $\approx 10^{13}:1$ at 25°. What is not quite so clear, however, is why the planar form should be so much the more stable (lower energy level) of the two. With three methyl groups as substituents, as in (42), the planar form would be expected to be somewhat more stable than (42a) on steric grounds—the reasonably bulky methyl groups are further apart in a planar as compared with a pyramidal conformation—but not by anything like the amount suggested above. It could be that the sp^2 hybridized state of carbon is just that much more stable than the sp^3 hybridized state, but there do not seem to be compelling reasons for that either. It has, however, been suggested that specific stabilization of the planar state could result from delocalization involving the hydrogen atoms of the substituent methyl groups:

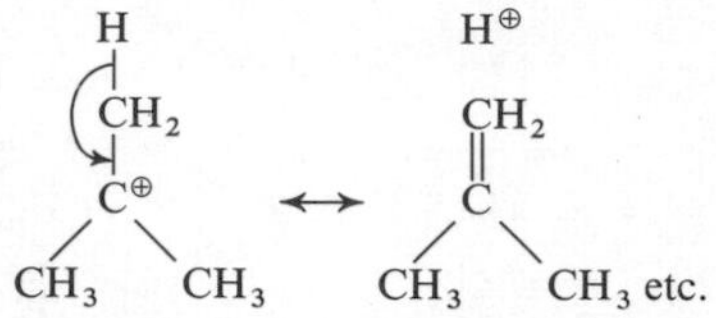

This is called *hyperconjugation*, and its operation would indeed be prevented—no p/sp^3 overlap possible—in the pyramidal conformation; there would thus be what is essentially steric inhibition of delocalization in both (43) and (42a). Some, though not unequivocal, support for the occurrence of hyperconjugation in (42) is provided by the observation that progressive replacement of the hydrogen atoms in 2-bromo-2-methylpropane (40) by deuterium results in slowing of the rate of solvolysis of the resultant deuterated bromide by ≈ 10 per cent per deuterium atom introduced. This reflects the expectation that the heavier deuterium atoms would be involved less successfully than hydrogen atoms in hyperconjugation, their increasing introduction would thus be expected progressively to destabilize the planar carbonium ion, and hence to slow the reaction.

DISPLACEMENT REACTIONS

(a) Nucleophilic. (i) S_N2—Inversion of configuration:

A necessary stereochemical concommitant of the concerted bimolecular mechanism (S_N2) of displacement (*cf.* p. 115) is that the carbon atom attached to the leaving group should undergo *inversion* of its configuration:

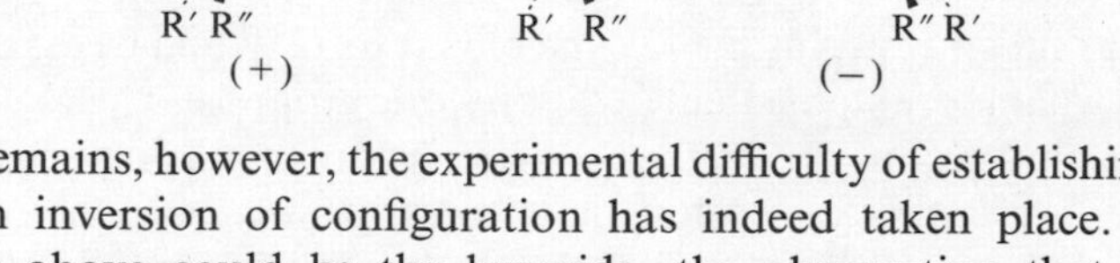

There remains, however, the experimental difficulty of establishing that such an inversion of configuration has indeed taken place. If the product above could be the bromide, the observation that it was laevorotatory, i.e. $(-)$, would of itself establish the occurrence of inversion, for it must then, by virtue of its direction of rotation, be the enantiomer of the initial substrate bromide which was dextrorotatory, i.e. (+). The product is in fact not, of course, the bromide but the corresponding alcohol, and the fact that the initial bromide and the product alcohol happen to exhibit opposite directions of rotation of plane polarized light is, unfortunately, no guide to whether their configuration is different or not: an independent establishment of their relative configuration is thus necessary.

This can be established unequivocally by dint of a series of inter-conversion reactions that need not trouble us here, and has, fortunately, been carried out already for a considerable number of the simpler dissymmetric species. Thus *dextrorotatory* methyl α-bromopropionate (44) has been shown to have the same configuration as *dextrorotatory* methyl α-methoxypropionate (45a):

$$\text{MeO}_2\text{C} \diagdown \quad\quad\quad \text{MeO}_2\text{C} \diagdown$$

$$\begin{array}{cc} \text{C—Br} & \text{C—OMe} \\ \text{Me} \diagup \vert \;\; (+) & \text{Me} \diagup \vert \;\; (+) \\ \quad\;\; \text{H} & \quad\;\; \text{H} \\ (44) & (45a) \end{array}$$

When (44) is reacted with $^{\ominus}$OMe in MeOH, the reaction is found to be second order and the product (45b) is found to be *laevorotatory*; we know therefore that inversion of configuration has indeed taken place in this second order reaction, as an S_N2 pathway would require:

$$\text{MeO}^{\ominus} + \begin{array}{c}\text{MeO}_2\text{C}\diagdown \\ \text{C—Br} \\ \text{Me}\diagup \; (+) \\ \text{H}\end{array} \longrightarrow \begin{array}{c}\text{CO}_2\text{Me} \\ \vert \\ \text{MeO}\overset{\delta-}{\cdots\cdots}\text{C}\overset{\delta-}{\cdots\cdots}\text{Br} \\ \diagup \;\; \diagdown \\ \text{Me} \;\; \text{H}\end{array} \longrightarrow \begin{array}{c}\text{CO}_2\text{Me} \\ \diagup \\ \text{MeO—C}\diagdown \;\; (-) \\ \text{H} \;\; \text{Me}\end{array} + \text{Br}^{\ominus}$$

$$\quad\quad\quad\quad (44) \quad\quad\quad\quad\quad\quad\quad\quad\quad\quad\quad (45b)$$

The reaction is, unexpectedly, found to be accompanied by a good deal of racemization, but this can be shown to arise from initial racemization of the starting material (44) before replacement of Br by OMe has occurred, i.e. it occurs quite independently of the actual displacement which is attended only by inversion.

An extremely simple and elegant demonstration that such bimolecular displacement is attended by inversion—one that avoids establishment of relative configuration of starting material and product, and other difficulties—involves the displacement of halide by the same, though isotopically labelled, halide ion; for example with dextrorotatory

2-iodooctane (46a) and sodium iodide labelled with ^{131}I, in acetone solution:

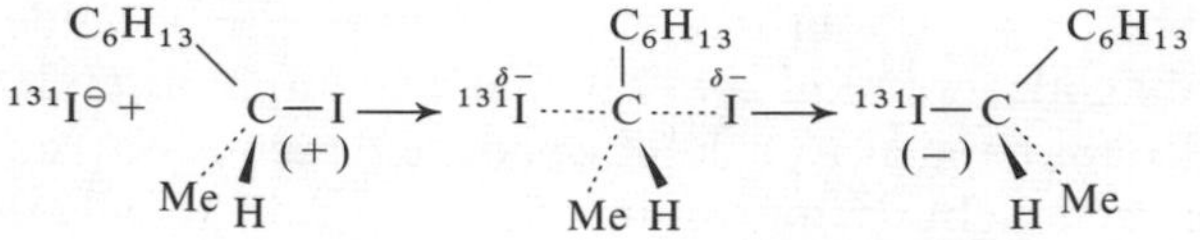

(46a)   (46b)

The rate of displacement can be followed by measuring the distribution of ^{131}I between 2-iodooctane and sodium iodide after various time intervals, i.e. the rate of radioactive exchange. The rate of inversion is obtained from measuring the rate of racemization (i.e. the rate of loss of optical activity, observed in the polarimeter); the former will then be half the latter, for each inversion of a molecule's configuration will result in the formation of a racemic pair of molecules in solution, a racemate being just a mixture of equal quantities of dextrorotatory and laevorotatory molecules.

The two measurements were, for convenience, made separately, not simultaneously on the same solution (this is irrelevant to the validity of the argument), and the rates then compared. It was found that the relative rates of displacement and inversion were 3.00 ± 0.25 and 2.88 ± 0.03, respectively. Their agreement is thus well within the experimental error, and it follows from this agreement that *each* act of bimolecular displacement must involve inversion of configuration.

(ii) S_N1—Racemization?:

The corresponding stereochemical concomitant of the unimolecular mechanism (S_N1) of displacement is that the carbon atom attached to the leaving group should undergo *racemization*

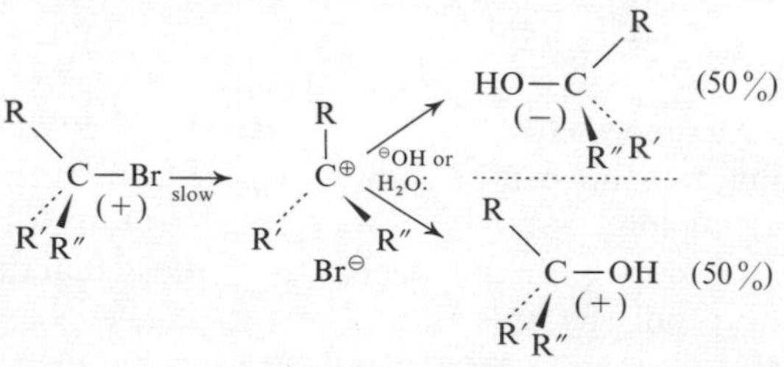

i.e. the planar carbonium ion moiety of the ion pair intermediate (47) would be expected to undergo subsequent fast nucleophilic attack, from either side with equal facility, to yield a 50/50 mixture of (+) and

(−) products. The experimental detection of whether such racemization actually takes place is, in theory, easy enough, but pure racemization is, in practice, observed very seldom: in reactions that are unquestionably first order, racemization is almost invariably accompanied, to a greater or lesser degree, by inversion. Racemization is thus very much less of a stereochemical requirement for the S_N1 pathway than inversion is for the S_N2; the reasons for this become apparent if we consider the ion pair intermediate (47), and its possible fate, a little more closely.

The relative proportions of racemization and inversion that occur are found to depend broadly on the structure of the initial halide, including its leaving group, and on the nature of the solvent—particularly on its ability as a potential nucleophile. This leads to the generalization that the more stable, and consequently—under given conditions— the longer lived, the carbonium ion in the intermediate (47) the greater the proportion of racemization that is observed. Thus the proportion of racemization tends to be greater with tertiary than with secondary halides, under otherwise comparable conditions, because tertiary halides yield carbonium ions that are, in general, more stable than those from secondary halides (simple primary halides do not generally undergo S_N1 displacements except under specialized conditions):

$$
\begin{array}{ccc}
\underset{CH_3 \quad CH_3 \text{ etc.}}{\overset{H\!-\!CH_2}{\underset{}{C^{\oplus}}}} & > & \underset{CH_3 \quad H \text{ etc.}}{\overset{H\!-\!CH_2}{\underset{}{C^{\oplus}}}}
\end{array}
$$

That the solvent must play an important role in S_N1 displacements is suggested by the fact that they are observed rarely, if at all, in the gas phase, and proceed only with difficulty in solvents of low polarity (e.g. hydrocarbons). By contrast, S_N1 displacements occur readily in solvents of high dielectric constant—the higher the dielectric constant the less energy is required for charge separation, i.e. ionization— particularly if these are effective solvating media, e.g. hydroxylic solvents that readily solvate both anions and cations. The sequence that involves the rate-limiting stage of an S_N1 process might be represented by:

$$
\overset{\delta+}{R}\!-\!X^{\delta-} \rightleftarrows \underset{(48)}{[R^{\oplus}X^{\ominus}]} \rightleftarrows \underset{(49)}{[R^{\oplus}\ X^{\ominus}]} \rightleftarrows \underset{(50)}{[R^{\oplus}] + [X^{\ominus}]}
$$

Here the solvent assists in the breaking of the R—X bond to form, initially, an intimate ion pair (48) in which the gegen ions are jointly surrounded, and stabilized, by an envelope of solvent molecules. This may be succeeded by a less close, but still discrete, ion pair (49) in which a solvent molecule(s) now separates the constituents. The final state is one in which the constituents of the ion pair are dissociated, each now being surrounded, and stabilized, by its own envelope of solvent molecules (50). Attack on $R^\oplus$ by a nucleophile present in the solution or, preferentially, by the more numerous and closely associated solvent molecules, if these are suitable, can take place at any stage along this sequence. The stabler the carbonium ion, however, the less readily it is likely to be attacked, and the further along the sequence we should thus expect it to persist.

Solvolysis of the halide (51) in 80 per cent aqueous acetone is found to lead to 98 per cent racemization and to only 2 per cent net inversion. This result reflects the formation of a relatively stable carbonium ion that is but little attacked by solvent in (48a), most of it persisting until it is symmetrically solvated in (49a)—or even perhaps until it is fully dissociated, corresponding to (50):

$$
\begin{array}{c}
C_6H_5 \\
\diagdown \\
\quad C-Cl \xrightarrow[\substack{80\% \text{ acetone}\\20\% \text{ water}}]{H_2O} \\
Me \diagup \\
\; H \quad (51)
\end{array}
$$

(48a) ⟶ (49a)

2%   49%   49%

$$
\begin{array}{cc}
& C_6H_5 \\
& \diagup \\
HO-C & \\
& \diagdown \\
+H^\oplus Cl^\ominus & H \quad Me
\end{array}
\qquad
\begin{array}{cc}
C_6H_5 & \\
\diagdown & \\
& C-OH \\
\diagup & \\
Me \; H & +H^\oplus Cl^\ominus
\end{array}
$$

What attack does take place in (48a) is, of course, going to lead to inversion through attack by H_2O: 'from the back', the 'front' side of the carbonium ion being shielded by the gegen-ion, $Cl^\ominus$.

By contrast solvolysis of the halide (52) is found to lead to only 17 per cent racemization and to no less than 83 per cent net inversion. This reflects the fact that (52) would be expected to form a less stable— and hence shorter lived—carbonium ion in (48b) than (51) did in (48a)—no stabilization by delocalization involving the π orbital system of a benzene nucleus in the former—and thus to undergo a much greater proportion of attack by solvent at the earlier, intimate ion pair stage:

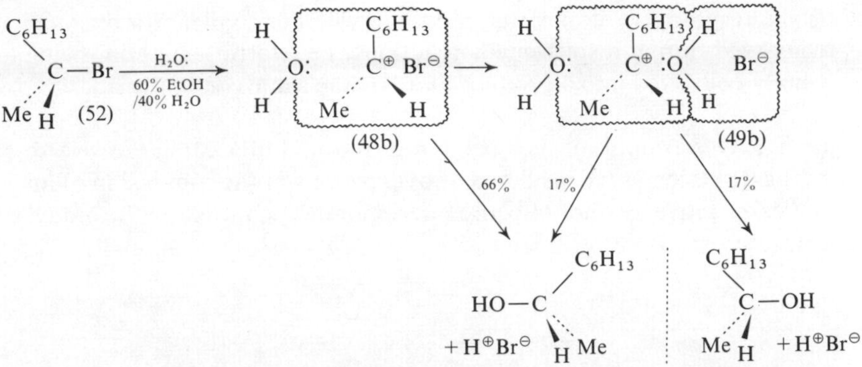

$$
\begin{array}{c}
C_6H_{13} \\
\diagdown \\
C-Br \\
\diagup \diagdown \\
Me H \quad (52)
\end{array}
\xrightarrow[\substack{60\% \text{ EtOH} \\ /40\% \text{ H}_2\text{O}}]{\text{H}_2\text{O:}}
$$

$$
\underbrace{\begin{array}{cc}
H & C_6H_{13} \\
\diagdown & | \\
O: & C^{\oplus} \; Br^{\ominus} \\
\diagup & \diagup \diagdown \\
H & Me \quad H
\end{array}}_{(48b)}
\longrightarrow
\underbrace{\begin{array}{ccc}
H & C_6H_{13} & H \\
\diagdown & | & \\
O: & C^{\oplus} \; :O & Br^{\ominus} \\
\diagup & \diagup \diagdown & \diagdown \\
H & Me \quad H & H
\end{array}}_{(49b)}
$$

66%   17%   17%

$$
\begin{array}{cc}
C_6H_{13} & \qquad C_6H_{13} \\
\diagup & \qquad | \\
HO-C & \qquad C-OH \\
\diagdown & \diagup \\
+H^{\oplus}Br^{\ominus} \quad H \; Me & Me \; H \quad +H^{\oplus}Br^{\ominus}
\end{array}
$$

Clearly the nature of the solvent or other nucleophile would also be expected to play an important part in deciding how far along the sequence attack takes place—the more powerful the nucleophile is, the sooner attack will take place on any given ion-pair, and the greater the proportion of inversion to racemization we should expect to see. Thus with the chloride (51) above, S_N1 displacement by H_2O: in 20 per cent water/80 per cent acetone leads as we have seen to 98 per cent racemization/2 per cent net inversion, whereas in the more strongly nucleophilic 40 per cent water/60 per cent acetone the corresponding figures are 95 per cent racemization/5 per cent net inversion, and in the even more strongly nucleophilic water alone, 82 per cent racemization/ 18 per cent net inversion: the proportionate shift with a less stable ion pair would, of course, have been more pronounced. It is important to emphasize that these solvolyses are all proceeding purely by the S_N1 pathway.

So far as the effect of the leaving group is concerned, we should expect the stronger the forces between the ions in the ion pair, the less likely is it to go from the intimate ion pair (48) to the more separated one (49), i.e. the 'tighter' the ion pair the greater the expected proportion of net inversion to racemization. Thus whereas the chloride (51) is solvolysed in water with 82 per cent racemization/18 per cent net inversion, the corresponding bromide—which would be expected to form a less 'tight' ion pair (bromine is less electronegative than chlorine)—is almost 100 per cent racemized on solvolysis in water.

(iii) Retention of configuration:

In the face of groups of displacement reactions attended by inversion of configuration (S_N2) and by racemization/inversion of configuration

(S_N1), respectively, it may well be asked whether configuration is ever conserved during displacement reactions, i.e. starting material being wholly converted into a product that has the same configuration.

(*a*) S_Ni: One example of such behaviour is found to occur in the reverse of the displacements considered above, namely in the conversion of an optically active alcohol into the corresponding chloride with thionyl chloride

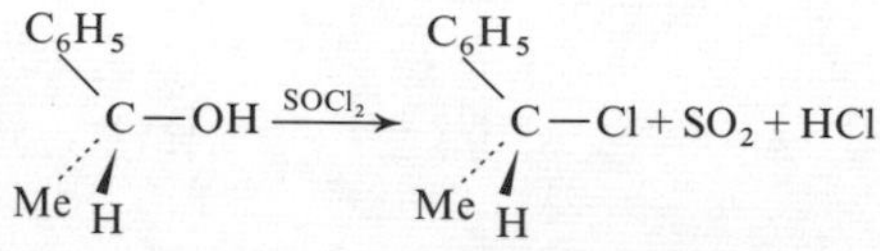

$$\underset{Me\;\;H}{\overset{C_6H_5}{C-OH}} \xrightarrow{SOCl_2} \underset{Me\;\;H}{\overset{C_6H_5}{C-Cl}} + SO_2 + HCl$$

clearly some special pathway must be operative to allow of the observed retention of configuration (*cf.* S_N2 above). With care, and under mild conditions, the chlorosulphite ester (53) can be isolated

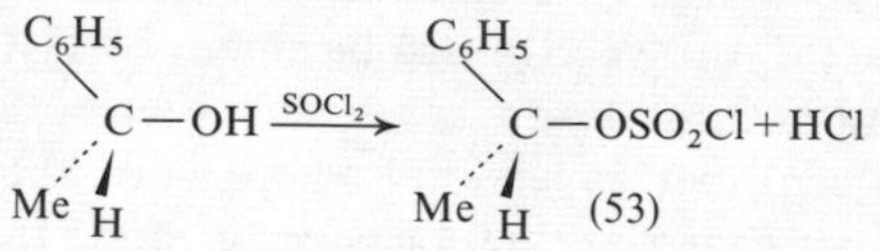

$$\underset{Me\;\;H}{\overset{C_6H_5}{C-OH}} \xrightarrow{SOCl_2} \underset{Me\;\;H}{\overset{C_6H_5}{C-OSO_2Cl}} + HCl \quad (53)$$

and shown to be a real intermediate by its subsequent conversion into the normal reaction products at a rate at least as fast as the overall reaction under comparable conditions (*cf.* p. 72). This conversion could be envisaged as proceeding by a concerted process such as

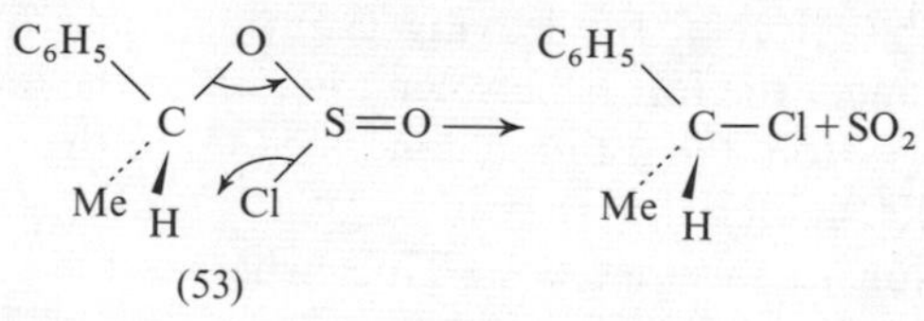

$$\underset{Me\;\;H\;\;Cl}{\overset{C_6H_5\quad O}{C}} S{=}O \longrightarrow \underset{Me\;\;H}{\overset{C_6H_5}{C-Cl}} + SO_2$$
$$(53)$$

in which the chlorosulphite ester (53) can take up a conformation such that the chlorine atom is well situated to attack the carbon atom 'from the front' as the latter's bond to oxygen is broken. It is found, however, that the rate of decomposition of (53) increases with the polarity of the solvent, and also with the stability of the carbonium ion in the ion pair (54) that could be formed from (53). It would thus appear that decomposition of (53) proceeds through an ion pair (54) rather than by the concerted process referred to above:

$$C_6H_5 \quad \underset{\underset{Me\,H}{\overset{}{C}}}{\overset{O}{\diagdown}} \overset{}{S}=O \longrightarrow \underset{Me\;H}{\overset{C_6H_5}{C^\oplus}} \overset{\ominus O}{\diagdown} S=O \xrightarrow{-SO_2} \underset{Me\;H}{\overset{C_6H_5}{C^\oplus}} Cl^\ominus \longrightarrow \underset{Me\,H}{\overset{C_6H_5}{C-Cl}}$$

(53)   (54)   (55)

Rapid decomposition of the chlorosulphite anion in (54) is believed
to take place, within the solvent envelope, to yield the new ion pair (55)
in which the chloride anion is associated with the 'front' of the carbonium ion; collapse of (55) results in the formation of the product chloride
with the same configuration as the starting material.

It is interesting that if the overall reaction is carried out in the
presence of pyridine, the resultant chloride is then found to have the
opposite configuration to that of the starting material. This is due to
HCl, formed from the initial reaction of the alcohol and thionyl
chloride, converting pyridine into its highly ionized hydrochloride:
thus a relatively high concentration of $Cl^\ominus$ is now available to effect an
S_N2 displacement (with inversion) on the chlorosulphite ester (53):

$$Cl^\ominus + \underset{Me\,H}{\overset{C_6H_5}{C-OSO_2Cl}} \longrightarrow \overset{C_6H_5}{\underset{Me\;H}{\overset{\delta-}{Cl}\cdots C\cdots\overset{\delta-}{SO_2Cl}}} \longrightarrow \underset{H\;Me}{\overset{C_6H_5}{Cl-C}} + SO_2 + Cl^\ominus$$

(53)

(b) Neighbouring group participation: Retention of configuration is
also observed in some cases where there is a group present in the
molecule that, by one kind of association or another, is able to preserve the 'back' side of the relevant carbon atom from attack, thus
leading to preferential attack from the 'front' by an external nucleophile. Thus α-bromopropionate anion (56) on treatment with low
concentrations of base is found, unlike methyl α-bromopropionate
(p. 127), to undergo displacement with retention of configuration:

$$\underset{Me\,H}{\overset{\ominus O_2C}{C-Br}} \xrightarrow[\text{(low conc.)}]{\ominus OH} \underset{Me\,H}{\overset{\ominus O_2C}{C-OH}}$$

(56)

The reaction is found—at low $[^\ominus OH]$—to follow a first order rate law,
i.e. the rate of reaction is independent of $[^\ominus OH]$, and is believed to

proceed via a zwitterion intermediate (57). The $CO_2{}^\ominus$ in (57) is able to interact with the carbonium ion carbon atom, necessarily from the 'back', thereby preventing attack by $^\ominus OH$ from that side; the non rate-limiting attack by $^\ominus OH$ must thus take place from the 'front' of (57) with consequent retention of configuration in the product (58):

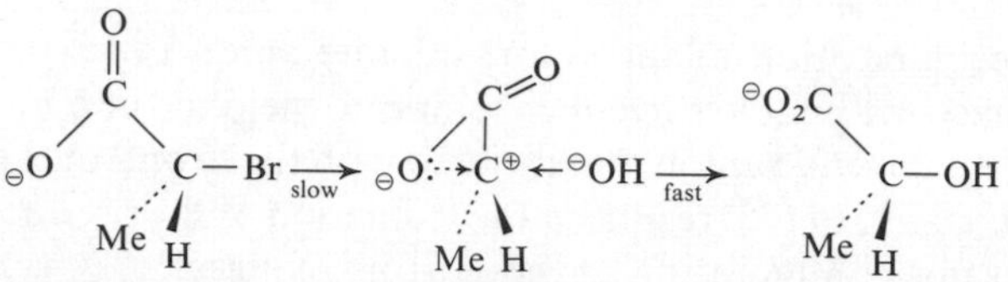

It has been argued whether or not an actual α-lactone intermediate (59) is ever formed

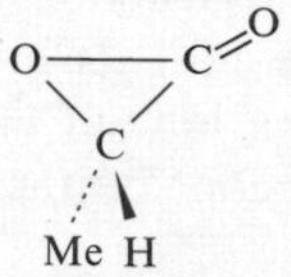

such a species would certainly be highly strained, and would be expected to undergo rapid ring-opening on attack by $^\ominus OH$. If such a species were formed, the observed 'retention' of configuration would in fact have been achieved by two successive inversions of configuration: the first by internal attack of $CO_2{}^\ominus$ on (56) to yield (59), and the second by $^\ominus OH$ on (59)—from the side away from oxygen—to yield the final product (58). Very recently it has been possible to isolate such an α-lactone (59a), which probably owes its relative stability to the bulky butyl groups inhibiting ready nucleophilic attack:

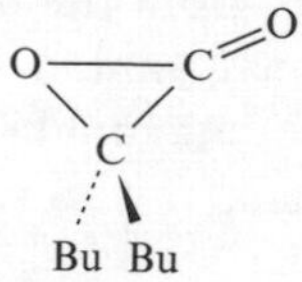

Whether or not they occur frequently as intermediates in such displacement reactions is not yet known, however. As $[^\ominus OH]$ is increased, progressively more product (60) is obtained that has a configuration opposite to that of the starting material, i.e. inversion is taking place due to increasing competition by a normal S_N2 displacement as $[^\ominus OH]$ rises:

$$HO^{\ominus} \;+\; \underset{\underset{\text{Me}}{\big|}\;\text{H}}{\overset{\overset{\ominus O_2C}{\diagdown}}{\diagup}} C - Br \longrightarrow \underset{\text{Me H}}{HO\cdots\cdots\overset{\overset{CO_2^{\ominus}}{\big|}}{C}\overset{\delta-}{}\cdots\cdots Br} \longrightarrow \underset{\underset{H}{\overset{\diagup}{\;}}\text{Me}}{HO - \overset{\overset{CO_2^{\ominus}}{\diagup}}{C}} \;\;+Br^{\ominus}$$

$$(56) \qquad\qquad\qquad\qquad\qquad\qquad (60)$$

The retention of configuration observed above when low $^{\ominus}OH$ concentrations were employed is said to arise from the operation of a *neighbouring group effect*—in this case through the agency of an oxygen atom of the carboxylate group; such effects are also observed in other displacement reactions where there is an atom, with an available electron pair, suitably situated in the molecule. Thus on reaction of an optically active (*erythro**) form of 3-bromobutan-2-ol (61) with HBr we expect to get attack from the 'back' by $Br^{\ominus}$ on the species (61a), in which the hydroxyl group has been reversibly protonated, inversion occurring at the carbon atom attacked, i.e. an ordinary S_N2 displacement, to yield the optically active dibromide (62):

$$(61) \qquad\qquad (61a) \qquad\qquad (62)$$

What is actually obtained, however, is the symmetrical, optically inactive (*meso*) form of the dibromide (63). This is believed to arise through the bromine atom, on the nearer carbon in (61a), acting as a neighbouring group by using an electron pair in a nucleophilic attack on the adjacent carbon to form the cyclic bromonium ion (64; *cf.* p. 85); the latter then undergoes attack—on either carbon atom—by

*In distinguishing the diastereoisomers of systems having two assymetric centres, the prefix *erythro* is applied to that form which, when represented in an eclipsed conformation, has the maximum number of like groups eclipsing each other. Considering the two

$$(61) \quad erythro \qquad\qquad\qquad\qquad (65) \quad threo$$

diastereoisomeric 3-bromobutan-2-ols (61) and (65), (61) is thus the *erythro* form. The other diastereoisomer (65) is referred to as the *threo* form, the names being derived from the diasteresisomeric four carbon aldose sugars erthyrose and threose, respectively.

$Br^\ominus$ to yield the alternative product molecules (63a and 63b):

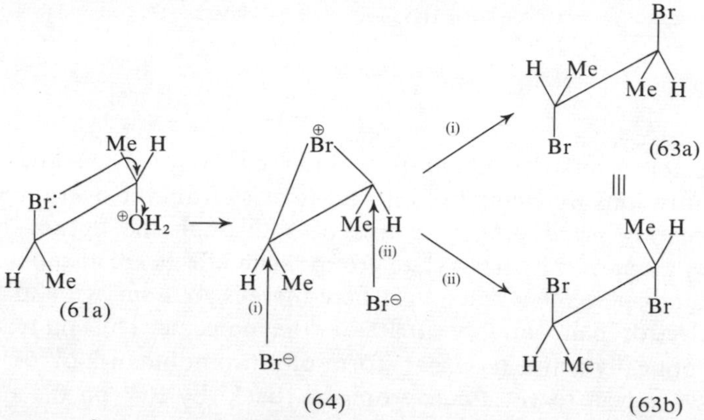

Me H

H Me
(61a)

$^\oplus OH_2$ $\longrightarrow$

$^\oplus Br$

Me H
(ii)

H   Me
(i)

$Br^\ominus$

(64)

(i)

$\longrightarrow$

Br

H   Me
Me   H

Br   (63a)

|||

Me   H

Br

Br

H   Me   (63b)

These are, of course, identical and symmetrical, i.e. the *meso* form is obtained.

Similarly, the protonated form (65a) of an optically active (*threo**) form of 3-bromobutan-2-ol (65) which would, by an S_N2 displacement, be expected to lead to the symmetrical, optically inactive *meso* product (66 ≡ 63a ≡ 63b),

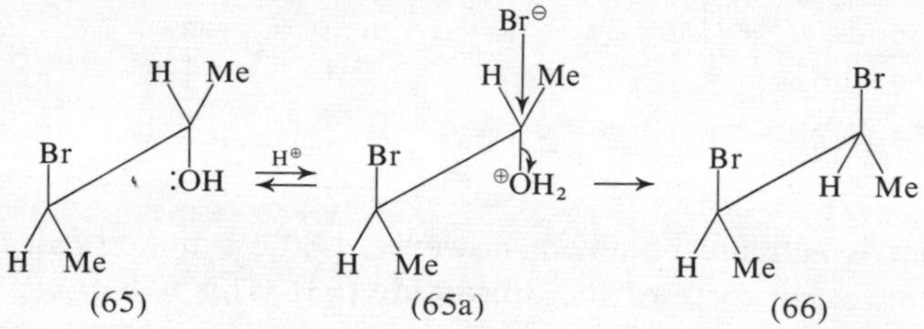

H   Me

Br

H   Me
(65)

:OH $\overset{H^\oplus}{\rightleftharpoons}$

$Br^\ominus$

H   Me

Br

H   Me
(65a)

$^\oplus OH_2$ $\longrightarrow$

Br

H   Me

Br   Me

H   Me
(66)

in fact yields equal quantities of (67) and (68), i.e. a racemic mixture, *via* the cyclic bromonium ion (69):

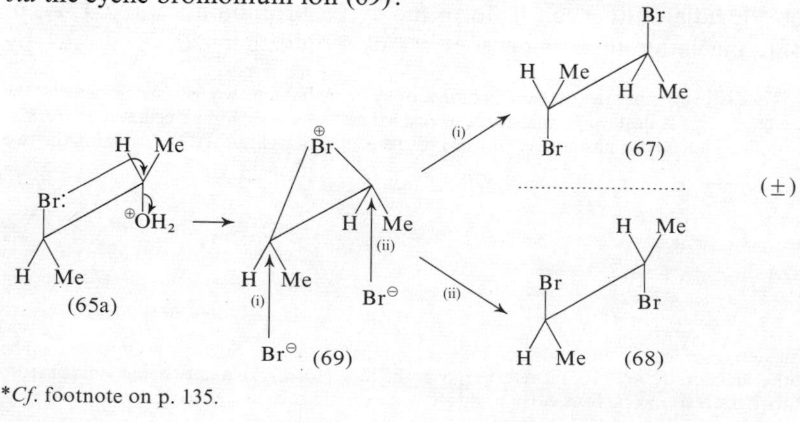

H   Me

Br:

H   Me
(65a)

$^\oplus OH_2$ $\longrightarrow$

$^\oplus Br$

H   Me
(ii)

$Br^\ominus$ (69)

(i)

(ii)

(i)

$\longrightarrow$

Br

H   Me
H   Me

Br   (67)

....   (±)

H   Me

Br

Br

H   Me   (68)

*Cf. footnote on p. 135.

The validity of such cyclic bromonium ion intermediates is heightened by the fact that they are also believed to figure in the addition of bromine to alkenes (p. 141); they have actually been detected, in a further context, by physical methods (p. 87). We have also seen an example (p. 81) in which the π orbital system of a benzene nucleus acts as a neighbouring group, though in this case without obvious stereochemical consequence.

A more general question arises in connection with neighbouring group participation: why can an internal nucleophile normally compete so favourably with an external one? There is evidence that the reason resides not so much in comparative energetics (ΔH^{+}) but in probability considerations (ΔS^{+}). If (65a), for example, were attacked by an external bromide ion a constraint would be imposed, in the transition state, on the relative (translational) motion of both species, and also on their orientation with respect to each other. When, however, the attack is by internal Br: only one species is involved (no translational or relative orientation terms arise), and the constraint imposed in the transition state relates only to restriction of rotation about a carbon-carbon bond such that a conformation is taken up which allows of attack by bromine's electron pair on the side of the adjacent carbon atom opposite to its protonated hydroxyl group. As translational entropy terms are much bigger than rotational ones, the entropy decrease ('ordering') in forming the transition state is considerably smaller for internal, rather than for external, attack and, given that the ΔH^{+} terms are much the same, the former will therefore be much the more rapid.

(b) Electrophilic

(i) Ortho/para ratios in aromatic substitution:

We have already referred to the bulk steric effect of groups already present in a benzene nucleus inhibiting attack on positions adjacent (*o*-) to them, and thus reducing the proportion of *o*-, compared with *p*-, product—the *o*-/*p*- ratio—in the resultant product mixture. In some cases, however, despite the presumed operation of such steric effects very large proportions of *o*- products are obtained—high *o*-/*p*- ratios— so much so that one is led to suspect the operation of some special effect.

Thus, if anisole (70) is nitrated with nitrating mixture (HNO_3/H_2SO_4) the product is found to contain 31 per cent *o*- and 67 per cent *p*- nitroanisoles (quite a normal distribution); if, however, the nitration is carried out with nitric acid in acetic anhydride, the product is now found to contain 71 per cent *o*- and 28 per cent *p*-nitroanisoles. Acyl nitrates, $RCO \cdot ONO_2$, are known to be formed from nitric acid and acid anhydrides and, though these probably do not usually function as nitrating agents, they could in this case yield 'incipient' $^{\oplus}NO_2$ through initial complexing (71) with the anisole:

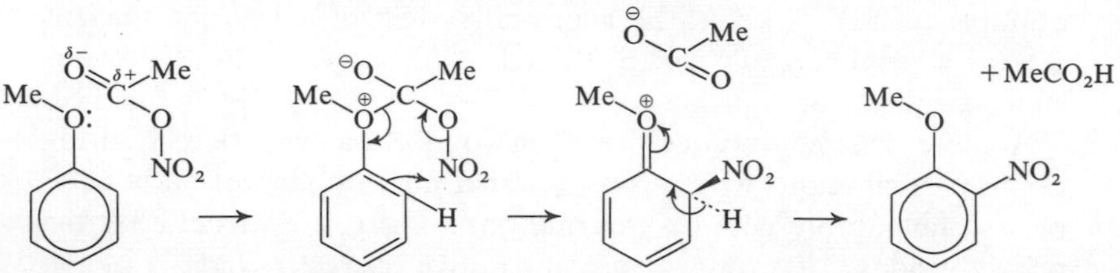

(70)   (71)

Nitration could then take place, *via* a 6-membered transition state, from a conformation in which the nitro group is directly adjacent to the *o*-position of the benzene nucleus.

A rather similar situation (probably involving a somewhat analogous pathway) arises with acetanilide where the product distribution is found to be 19 per cent *o*- and 79 per cent *p*- for nitration with nitrating mixture, compared with 68 per cent *o*- and 30 per cent *p*- for nitration with nitric acid in acetic anhydride. With methyl 2-phenylethyl ether (72) the corresponding values are found to be 32 per cent *o*- and 59 per cent *p*- with nitrating mixture, compared with 69 per cent *o*- and 28 per cent *p*- with N_2O_5 ($NO_2^{\oplus}NO_3^{\ominus}$) in methyl cyanide. In this latter case a different but analagous pathway is believed to be operative:

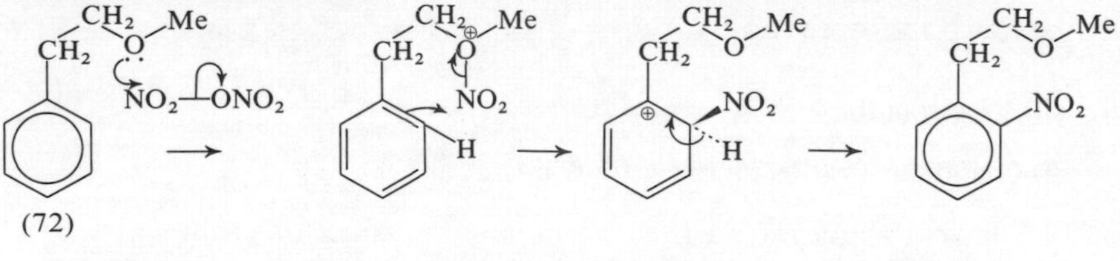

(72)

Such effects are not confined to nitration, however, and there is evidence that the formation of a very high proportion of the *o*- product from the reaction of formaldehyde with phenoxide ion (73) results from electron redistribution in a complex formed between the formaldehyde and the metal ion of the metal/phenoxide ion pair:

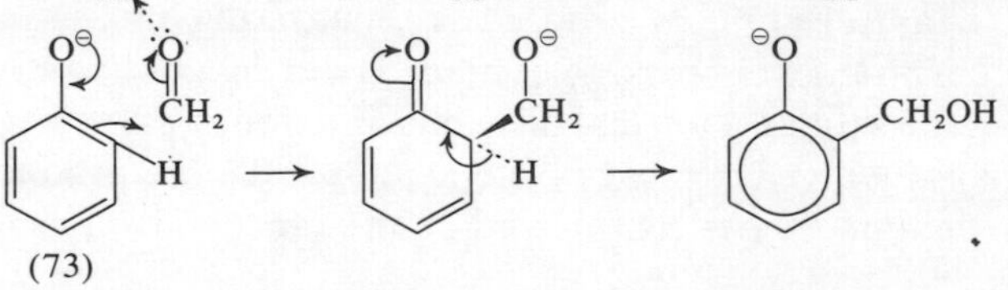

(73)

The above interactions can also be looked upon as examples of neighbouring group participation, albeit of a somewhat different kind from those we had considered above (p. 133).

(c) Radical

It is important to emphasize that radical displacement reactions such as the halogenation of an alkane are not direct displacements on carbon but involve initial abstraction of a hydrogen atom, e.g. in bromination:

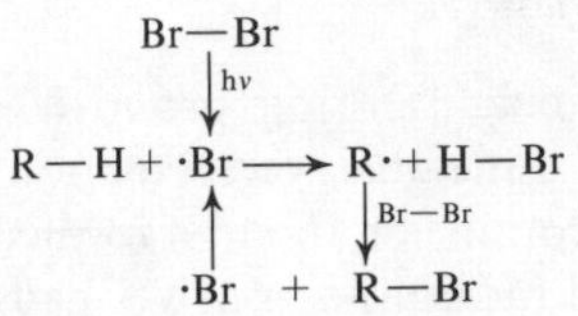

$$Br—Br$$
$$\downarrow h\nu$$
$$R—H + \cdot Br \longrightarrow R\cdot + H—Br$$

with $\cdot Br + R—Br$ and $Br—Br$ steps as shown.

What might be expected to happen when the attack is on an optically active alkane, $RR'R''CH$, will thus depend on the ultimate fate of the first formed radical, $RR'R''C\cdot$. Chlorination of the optically active halide (74) was found to result in the formation of a racemate:

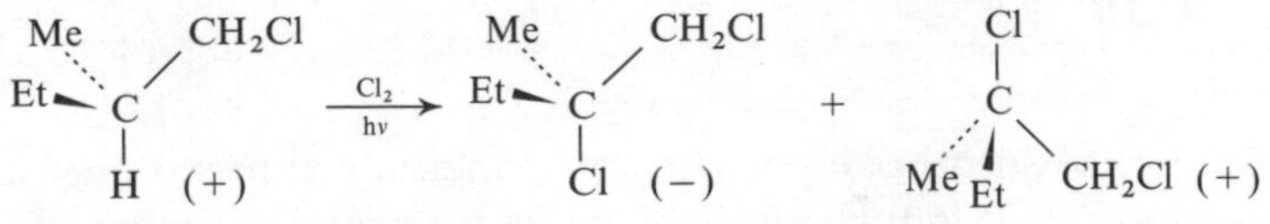

(74)   50%   50%

Subsequently the more selective bromination of the corresponding bromide (75)—or of (74) itself—was found to result in an optically active product having the same configuration as the starting material:

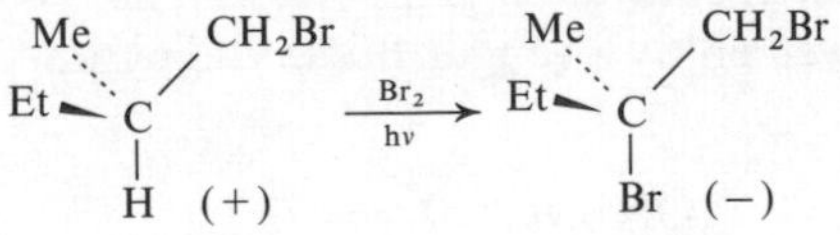

(75)

It has been suggested that this occurs *via* a cyclic bromonium radical (76)—bromine acting as a neighbouring group—with the result that subsequent attack can take place only from one side (*cf.* cyclic bromonium cations, p. 141):

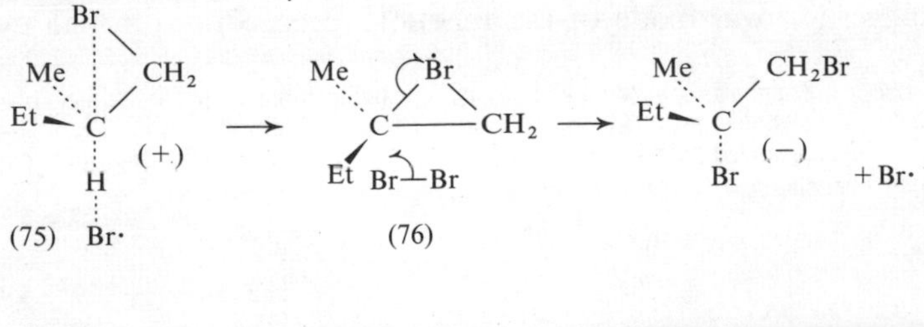

(75) Br·   (76)

There are some objections to the formation of cyclic bromonium radicals, however, and in any case the corresponding optically active cyano compound (77) has also been shown on bromination to yield an optically active product having the same configuration as the starting material:

$$
\underset{(77)}{\underset{\text{H} \quad (+)}{\overset{\text{Me} \qquad \text{CH}_2\text{CN}}{\text{Et}\smallsmile \text{C}}}} \quad \xrightarrow[h\nu]{\text{Br}_2} \quad \underset{\text{Br} \quad (-)}{\overset{\text{Me} \qquad \text{CH}_2\text{CN}}{\text{Et}\smallsmile \text{C}}}
$$

In this case no such neighbouring group effect can occur, and the simplest explanation for the observed retention of configuration would be that the attack of bromine on the first formed radical is so rapid that the initially pyramidal radical has not yet had time to collapse to the planar state:

$$
\underset{\substack{(77)\ \text{H} \\ \text{Br}\cdot}}{\overset{\text{Me} \qquad \text{CH}_2\text{CN}}{\text{Et}\smallsmile \text{C}}} \quad \xrightarrow{-\text{HBr}} \quad \underset{\text{Br}-\text{Br}}{\overset{\text{Me} \qquad \text{CH}_2\text{CN}}{\text{Et}\smallsmile \text{C}}} \quad \longrightarrow \quad \underset{\text{Br} \qquad +\text{Br}\cdot}{\overset{\text{Me} \qquad \text{CH}_2\text{CN}}{\text{Et}\smallsmile \text{C}}}
$$

In support of such finite persistence of the radical in pyramidal form it can be said that though physical evidence suggests a preferred planar configuration for simple alkyl radicals there is, in contrast to carbonium ions (p. 124), no apparent difficulty in forming radicals at bridgehead positions. The latter must retain a pyramidal configuration, and the much smaller difference in energy level between pyramidal and planar states for radicals, in contrast to carbonium ions, suggests that radicals in general may well enjoy a longer life in the pyramidal configuration.

ADDITION REACTIONS (C=C)

(a) Halogens. (i) Polar addition:

The electrophilic addition of both chlorine and bromine to simple acyclic alkenes is found to proceed with a high degree of TRANS* stereospecificity ($\approx$ 99 per cent); it was in order to explain this fact that the suggestion was made of the reactions proceeding through cyclic

---

*Where it is the *mode of addition* of X_2 that is being referred to—both Xs from the same side of the molecule: CIS; one X from each side of the molecule: TRANS—the term will be capitalised to distinguish it from reference to individual geometrical iso-merides where the *cis* and *trans* will be italicized.

halonium intermediates e.g. with *cis*- and *trans*-2-butenes (78 and 79, respectively):

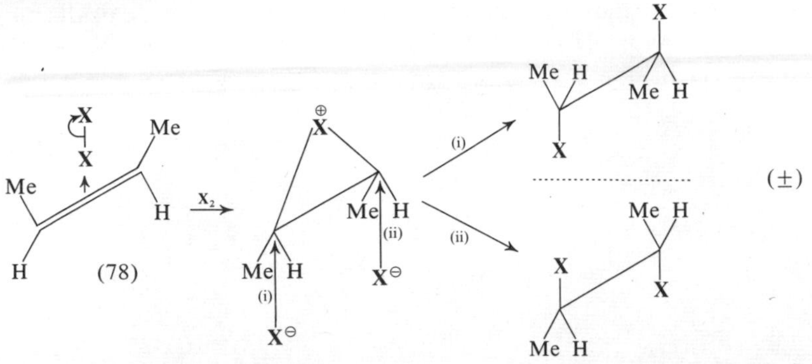

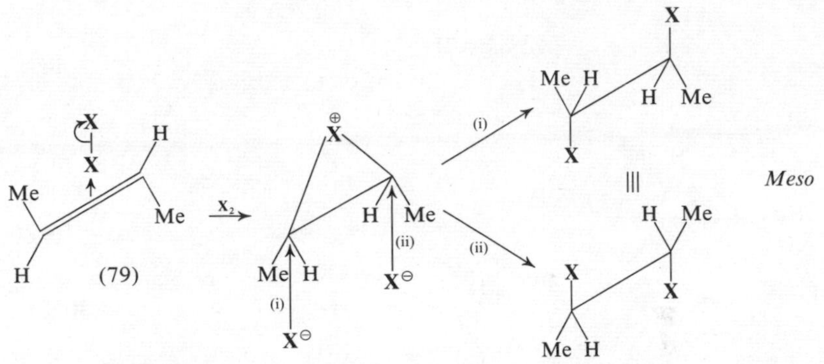

X
Me H
Me H
X

X
Me
X
Me
H $\xrightarrow{X_2}$
(i)
X⊕
Me H
(ii)
(i)
X
(ii)
(±)
Me H
X
(ii)
(ii)
X
X⊖
X
Me H
Me⟩H
(i)
X⊖

(78)

X
Me H
H Me
X

X
Me
X
Me
H $\xrightarrow{X_2}$
Me
(i)
X⊕
X
|||
Meso
H Me
(ii)
(i)
X
(ii)
H Me
(ii)
Me H
X⊖
X
X
H (79)
Me⟩H
(i)
X⊖
Me H

This purely *ad hoc* hypothesis has received considerable support from the ability of Cl and, particularly, Br to act as neighbouring groups (*cf.* p. 135), and especially from the direct detection of cyclic halonium ions by n.m.r. spectroscopy (p. 86). The kinetic laws for bromination are often found to be rather more complex than those for chlorination, and for the former $X^\ominus$ may sometimes be $Br_3{}^\ominus$ rather than simple $Br^\ominus$.

Alkenes that are capable of forming more highly stabilized carbonium ions are found to exhibit less stereospecific behaviour in the addition of halogens; thus *trans*-1-phenylpropene (82) and its *cis*-isomer are both found to yield 15–30 per cent of the products of CIS addition on reaction with bromine in non-polar solvents, and a higher proportion still in polar solvents. This is believed to arise through part of the reaction proceeding through the intimate ion pair (*cf.* p. 129) intermediates (80a and b)—the two different conformations arise by initial attack from above and below the planar alkene (82), respectively. These ion pair intermediates are in competition with the more familiar cyclic bromonium ion intermediates (81a and b)—arising from above and below attack, respectively—and are likely to compete the more effectively the more the carbonium ions can be stabilized (by delocalization over the π orbital system of the benzene nucleus, or other means), for no equivalent stabilization can operate in the cyclic bromonium ions:

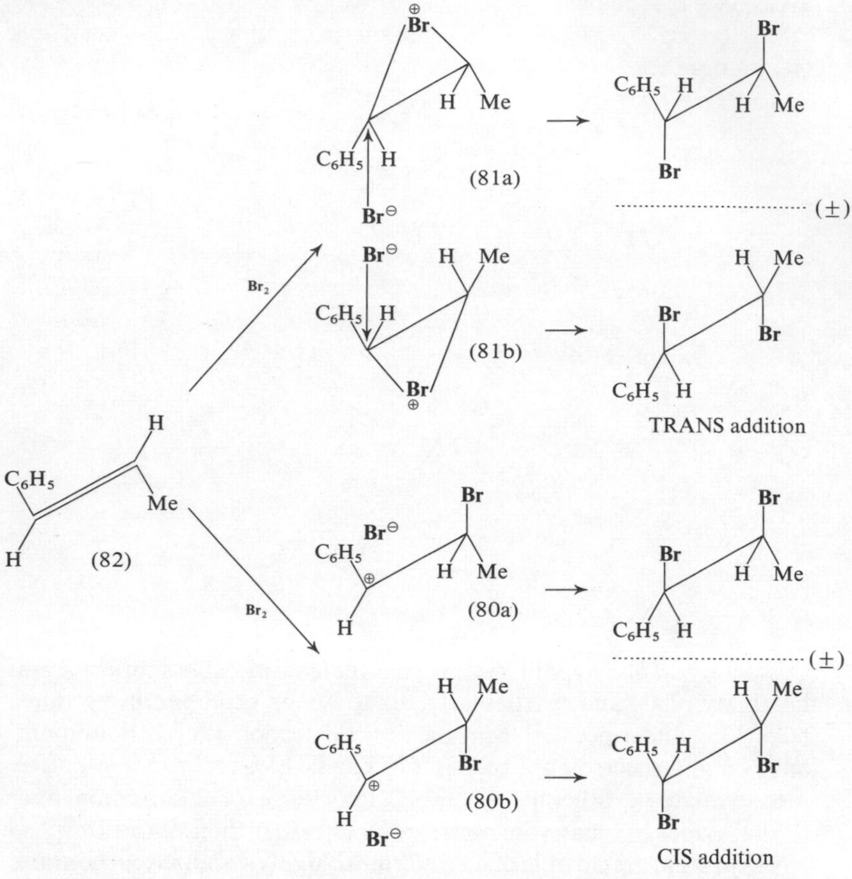

With XC_6H_4, where X is an electron-donating group (e.g. *p*-OMe), in place of C_6H_5 we should therefore expect to get a higher proportion of the product of CIS addition: this does indeed happen. The charge distribution in the cyclic bromonium ions (81a and b) will be unsymmetrical, and attack by $Br^\ominus$ will occur predominantly on the carbon atom shown. The two racemates obtained are, of course, different from each other as close scrutiny will show. Specific CIS addition will result from (80a and b) provided the intimate ion pair collapses to product rapidly; if, on the contrary, it enjoys a relatively long life before collapsing to product some rotation about the C_1/C_2 bond (originally the double bond) may take place, or the $Br^\ominus$ may have time to move round to the other side of the molecule: the products of both CIS and TRANS addition could then be obtained *via* (80a and b).

Increasing the polarity of the solvent would be expected progressively to stabilize the ion pairs (80a and b) by solvation, thereby making their formation (or, strictly, the formation of the respective transition states that precede them) less energy demanding; the proportion of the total reaction proceeding *via* the ion pair pathway (CIS addition) might thus be expected to increase with solvent polarity: as is indeed observed. For a given alkene, and under comparable reaction conditions, we should expect a higher proportion of TRANS (*via* cyclic halonium ion) addition with bromine than with chlorine, because bromine is the better bridging atom—more ready to share its electron pairs. This too is observed; thus acenaphthene (83) is found to yield largely the TRANS ($\pm$) addition product (84) with bromine (*via* the cyclic bromonium ion) but mainly the CIS (*meso*) addition product (85) with chlorine:

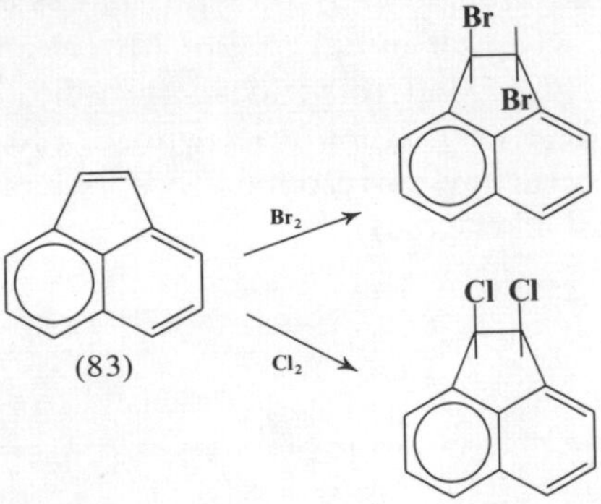

A complicating factor with such relatively rigid structures could, however, be that steric interaction between the two bulky bromine atoms in the product of CIS addition of bromine (*cf.* p. 149), and more particularly in the transition state that precedes it, could militate against its formation.

(ii) Radical addition:

In contrast to the above, relatively little is known about the detailed stereochemistry of radical-induced addition of halogens to simple alkenes (86); more is known about the addition of halogen hydracids (p. 147). There has, however, been some discussion about whether a bridged (87) or open chain (88) radical intermediate is involved:

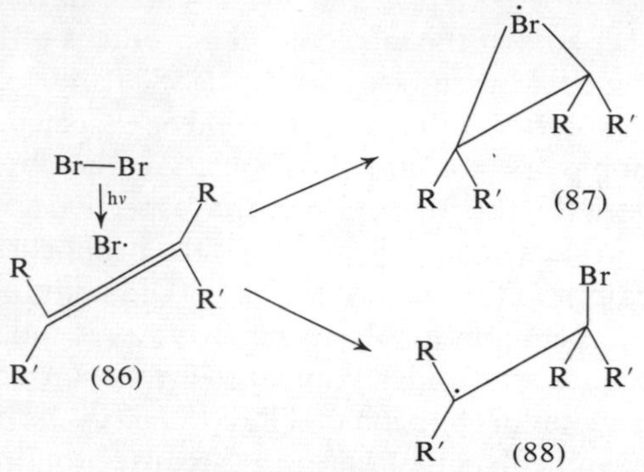

The former would lead to stereospecific TRANS addition (*cf.* addition *via* cyclic bromonium ions, p. 140), and the latter might also be expected to lead to preferred TRANS addition—attack on the radical (88) by a bulky bromine molecule is more likely to take place from the side opposite to the bromine atom already present—unless attack by Br_2 follows sufficiently slowly after initial radical formation as to allow rotation about the C_1/C_2 bond to yield some radicals having the conformation (89). These would then react with Br_2 to yield the di-bromide (90), whose structure corresponds to apparent overall CIS addition to the original alkene (86):

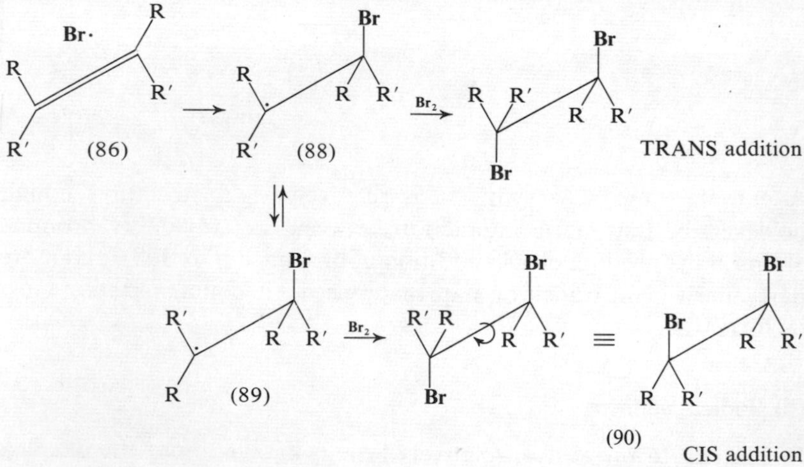

That such rotations about the C_1/C_2 bond do occur is demonstrated by the fact that small amounts of halogen, particularly bromine, will catalyse the interconversion of *cis-* and *trans-* alkene derivatives, and this has been shown to proceed *via* radical intermediates:

$$
\begin{array}{ccc}
\overset{\displaystyle H}{\underset{\displaystyle CO_2H}{\diagup\!\!\!\diagdown}}\!\!\!\diagdown CO_2H & \overset{Br_2}{\rightleftharpoons} & \overset{\displaystyle Br}{H\cdot\diagdown}\!\!\!\diagdown \overset{}{H\;\;CO_2H} \\
 & & \\
 & \updownarrow & \\
 & & \\
\overset{\displaystyle H}{CO_2H\diagup\!\!\!\diagdown}\!\!\!\diagdown CO_2H & \underset{-Br\cdot}{\rightleftharpoons} & \overset{\displaystyle Br}{CO_2H\,\cdot\diagdown}\!\!\!\diagdown \overset{}{H\;\;CO_2H}
\end{array}
$$

An equilibrium mixture is obtained in which the more stable, usually the *trans-*, alkene predominates; the fact that such isomerizations can occur, clearly prompts the exercise of caution in interpreting the stereochemistry of addition to an initial *cis-* or *trans-* alkene on the basis of an analysis of the stereochemistry of the products that are obtained.

The effect, on the first formed radical, of the length of time that elapses before subsequent attack by Br_2, or other chain transfer agent, is demonstrated in the radical-induced addition of thiols, RSH. These are known to carry on the reaction chain considerably more slowly than Br_2 or HBr and, in agreement with the predictions made above of the results of longer radical life, they are found to yield considerably greater amounts of the product derived from apparent CIS addition. One might also expect the extent of apparent CIS addition to be influenced by the concentration of a given transfer agent: the greater its concentration, the shorter the life of the initial radical and, consequently, the less the amount of the product of apparent CIS addition to be expected; this too has been observed.

A number of the above interpretative problems would be simplified by consideration of the stereochemistry of the radical-induced addition reactions of cyclic alkenes. A few such reactions have been studied but no clear picture has yet emerged, other than some apparent preference for TRANS addition. It is found, hardly surprisingly, that the local stereochemical features of the molecule undergoing attack play an important part in determining the overall reaction pathway.

(b) Halogen hydracids. (i) Polar addition:

There is no detailed information on the stereochemistry of addition of hydrogen halides to simple, acyclic non-conjugated alkenes. Attack is proton-initiated, in polar solvents at least, but it is not always possible to determine whether this proceeds *via* a simple carbonium ion (91) or a protonated double bond (92; a π complex: a cyclic σ complex, corresponding to a cyclic bromonium ion (p. 141), will not be formed as H, unlike Br, has no electron pair available):

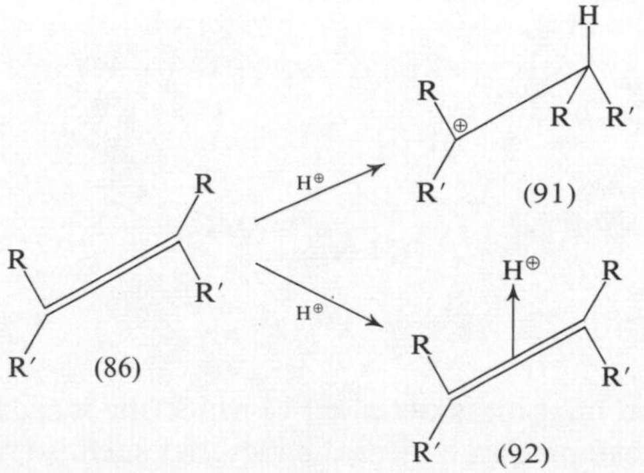

With alkenes that can form stabilized carbonium ions, addition almost certainly proceeds through (91) and, provided the intimate ion pair is not too long lived, we might expect it to collapse to yield largely the product of CIS addition. Thus polar addition of DBr to *cis*-1-phenyl-propene (93) in methylene dichloride is found to lead to (attack from only one side of the alkene molecule—above—is shown here):

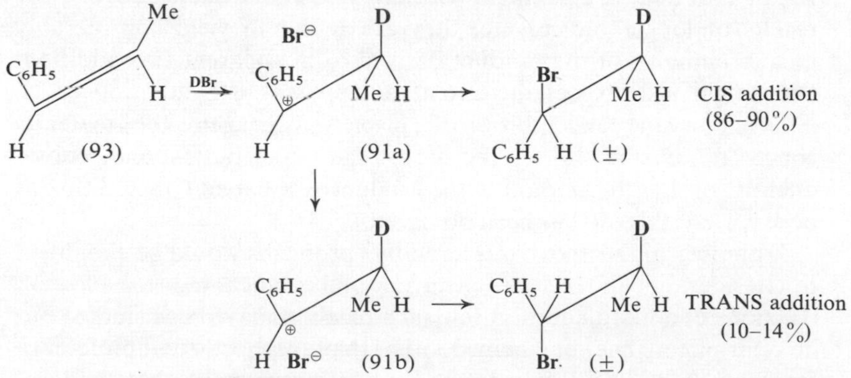

The product of TRANS addition is believed to arise through a certain amount of rearrangement of the original, intimate ion pair (91a ⟶

91b) before it collapses to product. In the polar addition of DBr to the isomeric *trans* 1 phenylpropene, the proportion of addition occurring by the two modes is found to be essentially the same: CIS (87–89 per cent) and TRANS (11–13 per cent). In general, such polar additions are found to become less stereospecific—progressively less product of CIS addition is formed—the more polar the solvent; this presumably arises from increasing stabilization by solvation, and hence longer life, of the ion pair (91) so that the chance of its isomerizing (91a ⟶ 91b) before collapsing to product are progressively increased.

Similar results are obtained with those cyclic alkenes that are also able to form stabilized carbonium ions in ion pairs, for example indene (94) and acenaphthene (95):

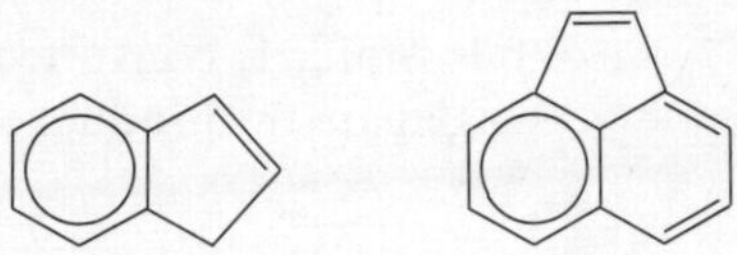

(94)   (95)

With simple cyclic alkenes that cannot form such stabilized carbonium ions some cases are known where addition is found to be predominantly TRANS; thus addition of HCl to 1,2-dimethylcyclopentene (96) in pentane solution is found to yield ≈ 100 per cent of the product of TRANS addition. The reason for this is not wholly clear but there is kinetic evidence in analagous cases that the transition state involves several molecules of HX, and a concerted reaction may be taking place in which the forming ions are stabilized, in the non-polar solvent, by further molecules of HX:

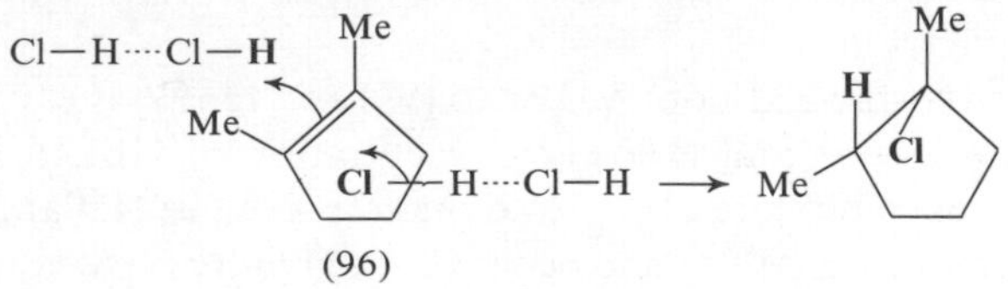

(96)

(ii) Radical addition:

The radical-induced addition of halogen hydracids—particularly HBr and DBr—to alkenes has been the subject of some considerable study; the results suggest TRANS addition as the preferred pathway unless particular circumstances militate against it. Thus at − 78° in liquid HBr, *cis*-2-bromobut-2-ene (97) is converted almost completely into the *meso*-dibromide (98): the product of TRANS addition:

Br

Br

Me

Me H   ——HBr——▶   Me Br   +Br·

Me H

Me   Br·   Br   H   (98)

|||

Me

H   Me H   Me H

Br·

Br   (97)   Me   Br   ——HBr——▶   H   +Br·

Me Br

Br   Me Br

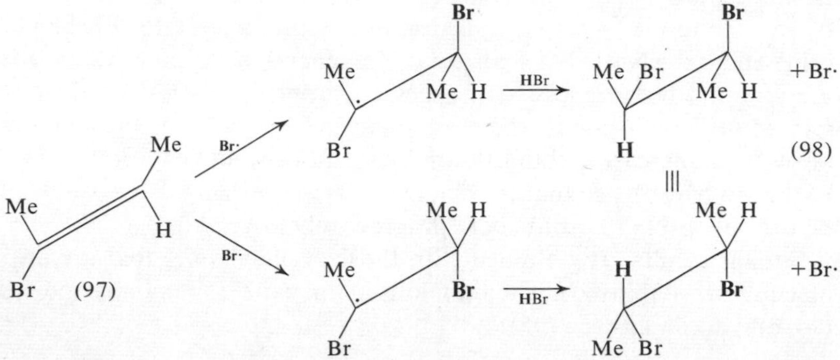

Trans-2-bromobut-2-ene (99) is similarly converted almost completely into the (±)-dibromide (100ab), again the product of TRANS addition:

Br

Br

Me

H Me   ——HBr——▶   Me Br   +Br·

H Me

H   Br·   Br   H   (100a)

·································· (±)

Me

Me   H Me   H Me

Br·

Br   (99)   Me   Br   ——HBr——▶   H   +Br·

Me Br

Br   Me Br   (100b)

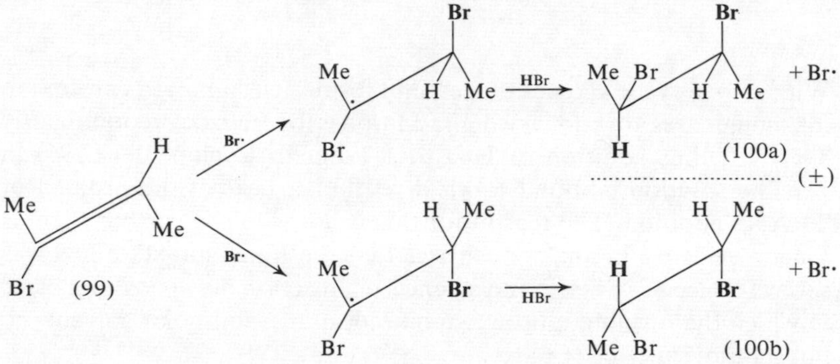

Similar preferred TRANS addition of DBr occurs with *cis*- and *trans*-2-butenes at −78°.

If, however, the addition of HBr to (97) and to (99) is carried out at room temperature, and in lower concentration of HBr, then each is found to yield a mixture of (±) and *meso*- products, (110ab) and (98), respectively. This is believed to occur through more rapid equilibration of the conformations of the radical intermediates, by rotation about the C_1/C_2 bond, at the higher temperature—as with the initial radical from (97)—

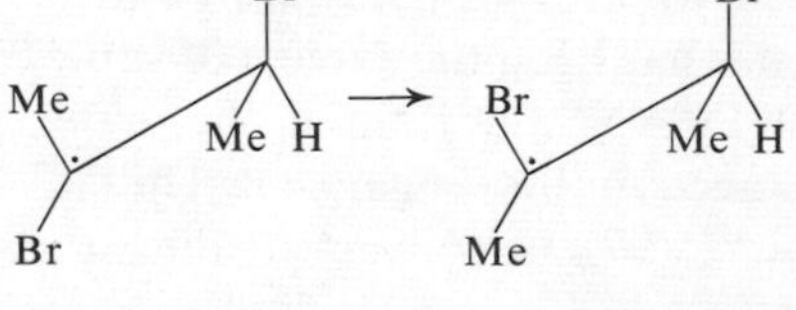

Br

Br

Me

Me H   ——▶   Br   Me H

Br   Me

assisted by the longer life of the radical at lower concentrations of HBr. In fact, the same mixture of products, 75 per cent ($\pm$) and 25 per cent *meso*, is obtained from either (97) or (99), reflecting the fact that the intermediate radicals have had the opportunity of coming to conformational equilibrium before the overall addition is completed by HBr attack to form the final products. It may be objected that this product mixture could arise through equilibration either of the original *cis*- and *trans*-alkenes (*cf.* p. 145) or of the final products, but this seems unlikely as the *meso*-dibromide (the minor product) is thermodynamically more stable than the racemate ($\pm$). As with the addition of halogens (p. 143) the preference for TRANS addition becomes less marked if the chain transfer is slow, e.g. with $CBrCl_3$ instead of HBr.

Any possible equilibration of the original alkenes can be ruled out by studying additions to cyclic alkenes: here again essentially TRANS addition is observed in the absence of disturbing features (see below). Thus 1-bromocyclohexene (101) undergoes TRANS addition to the extent of at least 99·7 per cent:

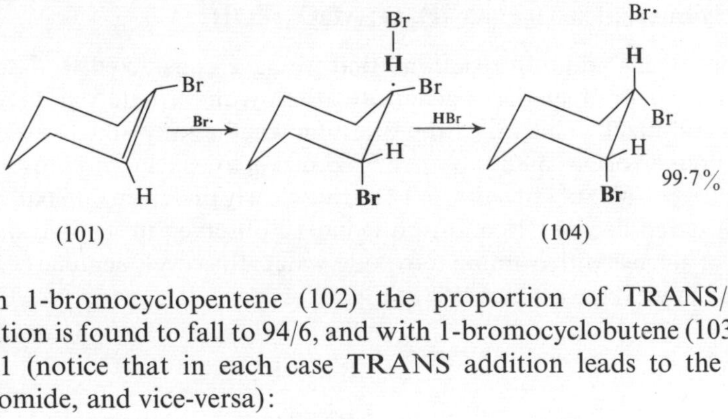

(101)   (104)

With 1-bromocyclopentene (102) the proportion of TRANS/CIS addition is found to fall to 94/6, and with 1-bromocyclobutene (103) to 79/21 (notice that in each case TRANS addition leads to the *cis*-dibromide, and vice-versa):

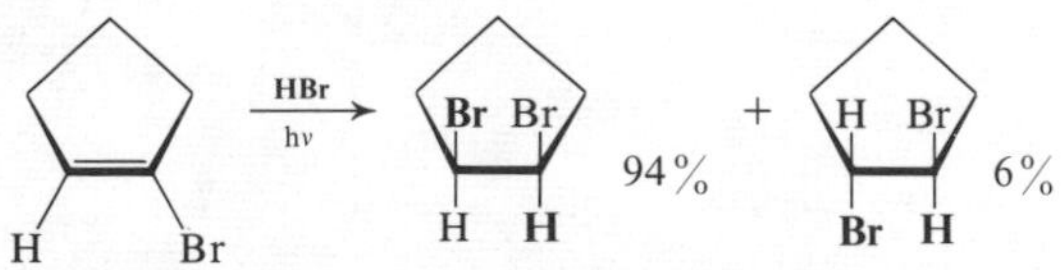

(105)

(102)

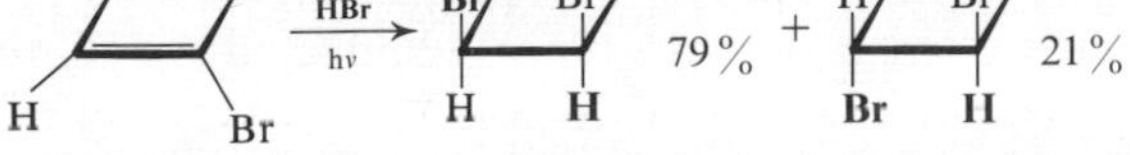

(103)   (106)

There is little doubt that the decreasing proportion of the product of TRANS addition, in the series 104 > 105 > 106, stems from increasing crowding of the two large bromine atoms in the *cis* configuration which raises the energy level of the corresponding transition state, thereby reducing its rate of formation. Any such steric strain can be avoided completely in the formation of (104), avoided to some extent by puckering in the formation of (105), but avoided considerably less well in the formation of (106).

Suggestions have been made (*cf.* p. 144) of the intervention of cyclic bromonium radicals to account for the observed degree of TRANS stereospecificity, but there seems no compelling reason to favour these over preferred attack by HBr on the side of the radical away from the Br atom that has already been bonded; it is, however, not possible to rule out entirely some interaction (short of bonding) between the lone pair on this bromine atom and the radical carbon atom.

(c) Hydroxylation. (i) OsO_4/H_2O; $MnO_4^{\ominus}/^{\ominus}OH$:

In the alkene addition reactions that we have considered to date the product or products have generally arisen from a preferred TRANS mode of attack. This raises the question whether any alkene addition reactions do occur, in the absence of special predisposing circumstances (e.g. the collapse of ion pairs, p. 141), by a clearly preferred CIS pathway. Such stereospecific CIS addition is indeed observed in the hydroxylation of alkenes with osmium tetroxide/water; thus cyclopentene (107) is found to yield the *cis*-diol (108) only:

$$\text{(107)} \xrightarrow[\text{H}_2\text{O}]{\text{OsO}_4} \text{(108)} + \text{H}_2\text{OsO}_4$$

(107)   (108)

That the product is the *cis*-1,2-diol may be confirmed by showing that it is optically inactive, i.e. a symmetrical *meso* form (the *trans*-1,2-diol is dissymmetric, and is thus obtained as a racemate that may be resolved), and also from the fact that its i.r. spectrum exhibits internal (intra-molecular) hydrogen bonding (108a), unaffected on dilution of the solution: a recourse that is not open to the corresponding *trans*-diol (109):

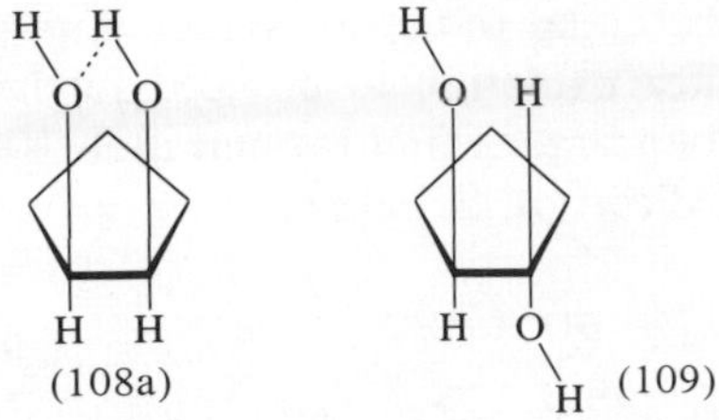

(108a)   (109)

The reason for the CIS stereospecificity of this addition was revealed when it proved possible to isolate a cyclic osmic ester intermediate (110)—which possible bond angles and distances require to have the *cis*-configuration—and to show that this could be decomposed (Os—O bond fission only) to the *cis*-diol (108) at a rate at least as fast as the overall reaction under comparable conditions:

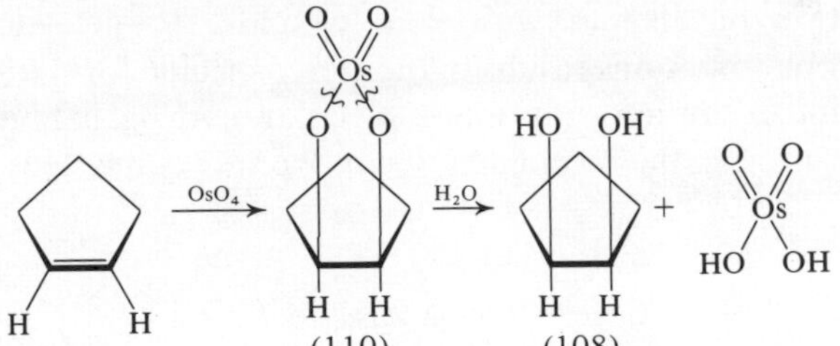

(107)   (110)   (108)

The more familiar alkaline permanganate—the classical reagent for the hydroxylation of alkenes—also leads, under mild conditions, to the *cis*-diol, and though it has not here proved possible to isolate a cyclic permanganic ester (111) corresponding to the cyclic osmic ester (110) above, the reaction is believed to proceed through such a species. Some supporting evidence is provided by the fact that carrying out the hydroxylation with ^{18}O labelled permanganate ($Mn^{18}O_4^{\ominus}$) is found to yield the *cis*-diol (108b) in which both the oxygen atoms introduced are ^{18}O labelled; they must therefore have been provided by the permanganate, as a cyclic permanganic ester (111) would require, and could not have come from the aqueous solvent:

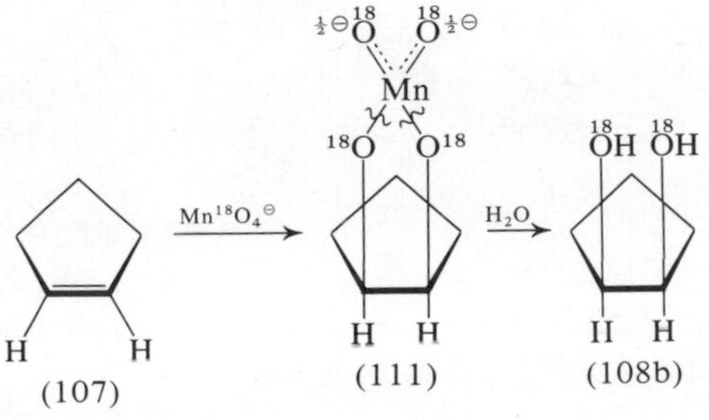

(107)   (111)   (108b)

Care has to be taken in the use of alkaline permanganate as it is extremely easy for oxidation to proceed further; fission of the 1,2-carbon/carbon bond in the first-formed diol (108) then takes place to yield the acyclic dicarboxylic acid (112):

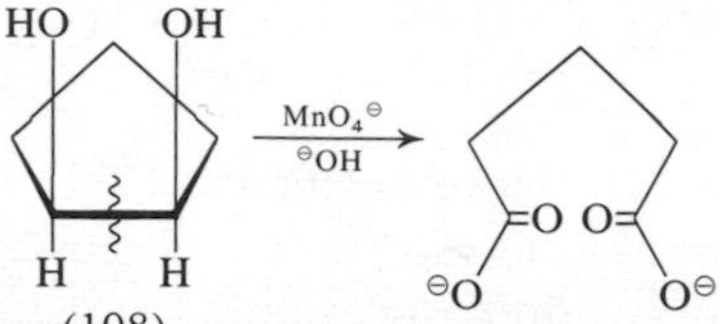

(108)   (112)

(ii) Peroxy acids:

Reaction of alkenes with peroxy acids, $R-\overset{\overset{\displaystyle O}{\|}}{C}-O-OH$, under vigorous conditions results in the stereospecific formation of the *trans*-1,2-diol exhibiting only intermolecular hydrogen bonding in its i.r. spectrum, e.g. the racemate (109ab) from cyclopentene (107):

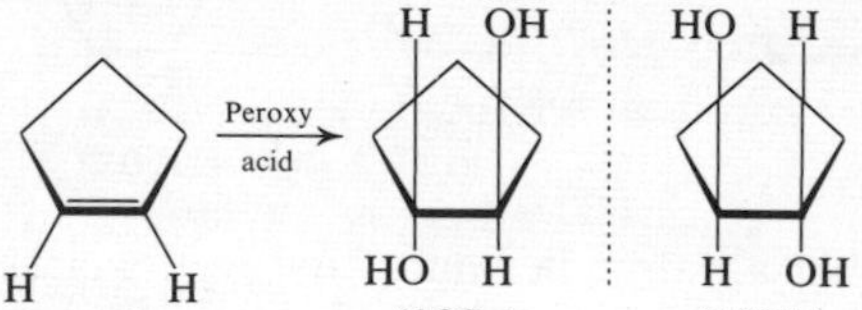

(107)   (109a)   (109b)

The key to the TRANS stereospecificity of the hydroxylation is revealed by the isolation, under milder—indeed normal—conditions, of the intermediate (113), a *cis*-epoxide:

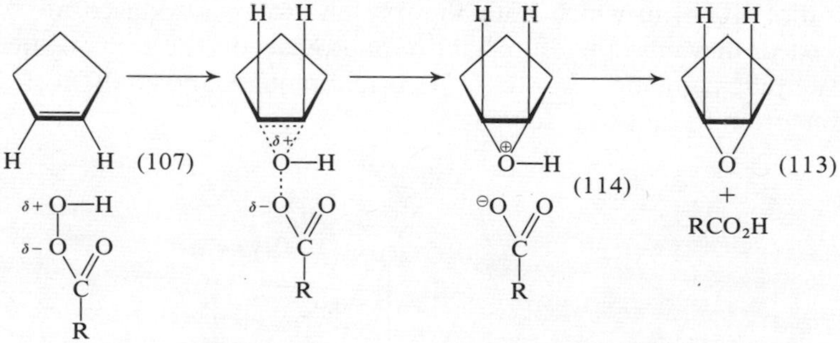

This three-membered epoxide ring may then be cleaved readily, by acid or base, to yield the *trans* ($\pm$)-diol (109ab):

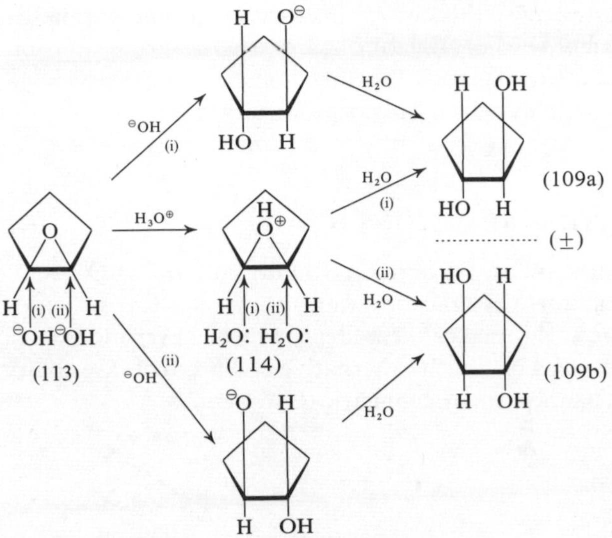

H   O⊖

⊖OH (i)   HO   H

H₂O → H   OH

(109a)

H₂O (i)   HO   H

·················· (±)

H₃O⊕ →   H O⊕

O

H (i)(ii) H
⊖OH⊖OH
(113)

⊖OH (ii)

H (i)(ii) H
H₂O: H₂O:
(114)

(ii) H₂O → HO   H

(109b)

H   OH

⊖O   H

H₂O

H   OH

In either case S_N2 attack takes place on carbon—with inversion of configuration—from the side of the molecule opposite to the epoxide oxygen atom, attack on one carbon leading to enantiomer (109a) and on the other to enantiomer (109b). Attack is statistically equally likely on either carbon atom in the symmetrical epoxide, and the result is a 50:50 mixture of (109a) and (109b), i.e. the *trans*-racemate (109ab).

If the acid, RCO₂H, resulting from the decomposition of the initial peroxy acid is sufficiently strong, the epoxide will remain in the protonated form (114) and may then undergo hydrolytic cleavage by water *in situ*. In the case of peroxyformic acid, the formate anion, HCO₂⊖, is sufficiently nucleophilic to attack (114), and the end-product of the reaction is thus the (±) *trans*-monoformate ester (115ab);

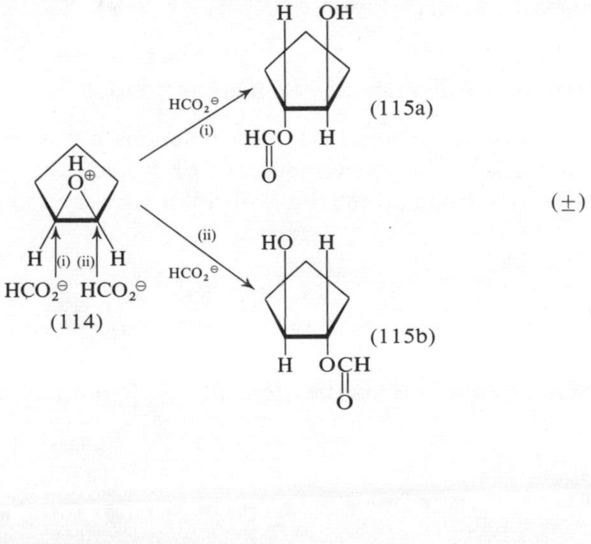

H   OH

HCO₂⊖ (i) →   HCO   H
                   ‖
                   O

(115a)

(±)

H O⊕

H (i)(ii) H
HCO₂⊖ HCO₂⊖
(114)

(ii) HCO₂⊖ →   HO   H

H   OCH
         ‖
         O

(115b)

this ester must then be hydrolysed with aqueous base to yield the racemic *trans*-diol (109ab).

These two methods together constitute an extremely useful, antithetic pair by which stereospecifically pure *cis*- and *trans*-1,2-diols, respectively, may be synthesized from alkenes at will.

ADDITION REACTIONS (C=O)

The question of CIS or TRANS addition to C=O does not of course arise as there is ready rotation about the C—O bond in the endproducts; this makes them identical whichever mode of addition they were formed by, i.e. the alternative end products are conformers rather than enantiomers or geometrical isomerides:

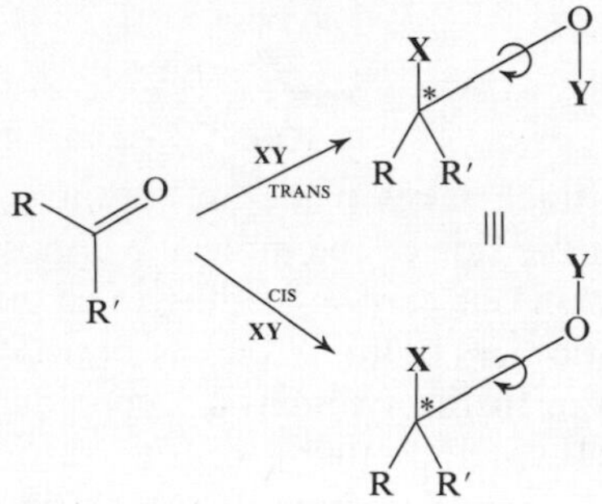

There are, however, two points of general stereochemical interest in connection with this reaction: (*a*) the bulk effect of R and R′ on the rate (and equilibrium position) of the addition, and (*b*) the effect of the non-symmetrical environment, when R or R′ is dissymmetric, on the proportions of enantiomers that are obtained from the generation of the new dissymmetric centre (*above).

(a) Bulk effects on reaction rate and equilibrium position

The great majority of carbonyl addition reactions are nucleophilic, arising from attack of the nucleophile, Y: or $Y^{\ominus}$, on the positively polarized carbonyl carbon atom; the X of XY is very commonly H:

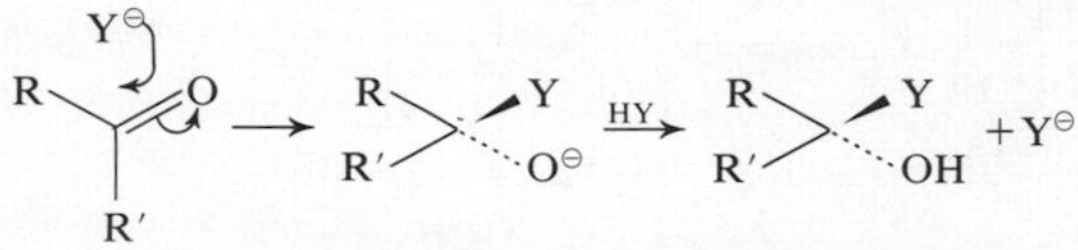

In the course of these additions the originally sp^2 hybridized carbonyl carbon atom becomes sp^3 hybridized in the adduct; the effect of this is to move R and R' closer together in the adduct (R —C—R' angle $\approx 109°$) than they were in the original carbonyl compound (R—C—R' angle $\approx 120°$). Any resultant steric interaction is of relatively little significance while R and R' are small, but as they increase in size their effect becomes more pronounced, progressively destabilizing the adduct with respect to the starting material. As a number of the nucleophilic addition reactions of carbonyl compounds come to equilibrium, rather than going substantially to completion, this is reflected in progressively smaller equilibrium constants, for example in the addition of bisulphite ion at 0°:

$$
\begin{array}{cc}
\ce{\underset{CH_3}{\overset{Me}{\diagdown}}C=O} & \ce{\underset{MeCH_2}{\overset{Me}{\diagdown}}C=O} \\
7{\cdot}9 \times 10^2 & 1{\cdot}58 \times 10^2
\end{array}
$$

It should be emphasized that electronic effects, i.e. electron-donation/ -withdrawal, also influence K but their effect in the above case is probably small. The size of the attacking nucleophile is also of significance in that, if large enough, it too can contribute to crowding, and consequent increase in energy level, in the adduct. Thus there is found to be very little difference between the equilibrium constants of the above ketones in cyanohydrin formation where the nucleophile is relatively small, and is also linear.

An interesting steric effect, this time stabilizing the adduct (117) with respect to the starting material, is seen with cyclopropanone (116) where the required C—C—C bond angle of $\approx 60°$ results in the original ketone (normal C—C—C bond angle $\approx 120°$) being more strained than the adduct (normal C—C—C bond angle $\approx 109°$). One result of this is that cyclopropanone's equilibrium with water in aqueous solution lies far over in favour of the hydrate (117) which may, indeed, be readily isolated:

$$
\ce{\overset{O^{\delta-}}{\underset{\delta+}{\triangle}}} \xrightarrow[\ce{<-}]{\ce{H_2O:}} \ce{\overset{HO\quad OH}{\triangle}}
$$

$$(116) \qquad\qquad (117)$$

It is found that steric effects also influence the rates of nucleophilic addition reactions, often in much the same general sequence as they do equilibrium constants, reflecting—as might perhaps have been expected—the fact that the transition state resembles the adduct more closely than it does the starting material. Thus the following relative reaction rates are observed for addition of bisulphite to a series of ketones in which variation in electronic effects is expected to be small:

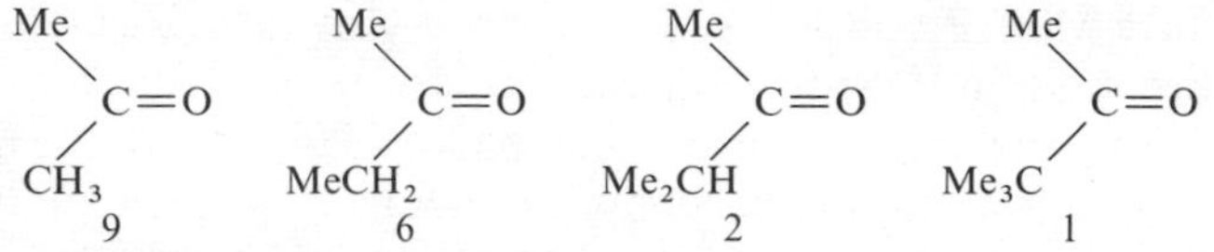

Me\
  C=O   CH$_3$ **9**

Me\
  C=O   MeCH$_2$ **6**

Me\
  C=O   Me$_2$CH   **2**

Me\
  C=O   Me$_3$C   **1**

(b) Cram's rule

In nucleophilic addition to a carbonyl compound a dissymmetric centre (*) is introduced but, provided R and R′ are not themselves dissymmetric, initial attack will be equally likely (statistically) from one side or the other of the essentially planar molecule

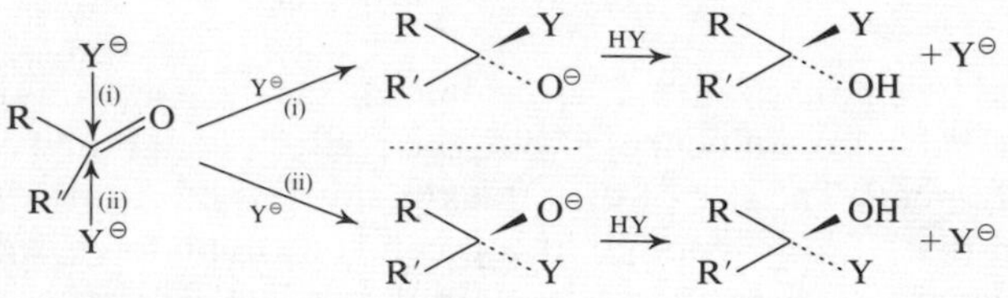

and a 50:50 mixture of the two enantiomers, i.e. a racemate ($\pm$) will result. If, however, the original carbonyl compound already possesses a dissymmetric centre—and particularly if it is α—to the carbonyl group—attack from one side of the carbonyl carbon atom will no longer be equally likely (statistically) as from the other. The alternative diastereoisomeric adducts (119 and 120; enantiomeric with respect to the new dissymmetric centre * that has been introduced) will no longer be produced in equal amounts:

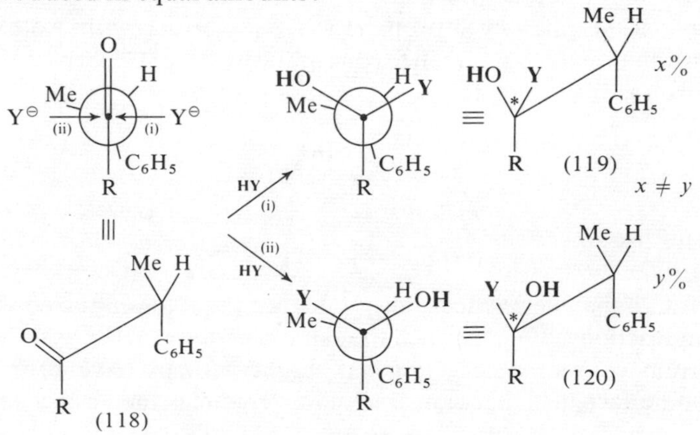

The question then arises whether it is possible, from knowledge of the detailed structure of the initial carbonyl compound, to predict which diastereoisomeric adduct is likely to predominate in the product. Such prediction is made possible through Cram's rule, which may be stated as follows: 'That diastereomer will predominate which would be formed by the approach of the entering group (nucleophile) from the *least hindered side* of the carbonyl carbon atom, when the rotational conformation of the $C_1 — C_2$ (carbonyl carbon/α-carbon) bond is such that the $C=O$ is flanked by the two least bulky groups attached to the adjacent (α-) dissymmetric centre.' The above ketone (118) is thus depicted in the required rotational conformation (Me and H are less bulky than C_6H_5) and, as H is less bulky than Me, we should predict attack (i) to predominate over (ii): we should thus expect more of (119) than of (120) in the product, i.e. $x > y$.

In the reaction of the Grignard reagent, methyl magnesium iodide, on 2-phenylpropanal (118a, i.e. 118, $R=H$)

Me H

HO   H Me   HO Me
Me—⟨   ⟩Me   ≡   ⟨*⟩ C_6H_5 *erythro*
          C_6H_5 H   (119a)
            H

O   (i)
Me ‖ H   MeMgI   (a) MeMgI   Me H
MeMgI —(ii)⟨⟩(i)—   (b) H⊕/H₂O
   C_6H_5 (ii)   Me   H OH   Me OH
   H   Me—⟨   ⟩   ≡   ⟨*⟩ C_6H_5 *threo*
                              C_6H_5 H   (120a)
  (118a)   H

the prediction would be that (119a), the *erythro* (*cf.* p. 135) product derived from the preferred, less hindered attack (i), would predominate: in practice the *erythro*: *threo* ratio is found to be 2:1. Reaction of the same Grignard reagent on the aldehyde in which Me has been replaced by the rather bulkier Et results in the observed *erythro*: *threo* ratio increasing to 2·5:1, while reaction of the very much bulkier Grignard reagent, C_6H_5MgBr, on the aldehyde (118a) results in the ratio of (i):(ii) attack going up to >4:1.

The rule has been found to make successful predictions in a very considerable number of cases, so much so that it has been used to establish the configuration of several products where this was pre-viously unknown. There are, however, a number of points about its use that need to be borne in mind; thus it does not usually apply to cyclic compounds, where other factors such as ring strain, preferred con-formations etc., have also to be considered. It also holds only for reactions that are kinetically controlled: if the products are able to attain equilibrium with each other under the conditions of the reaction, e.g. reversible addition, etc., then the product ratio will be determined by their relative thermodynamic stabilities under the conditions of the reaction.

The reason for specifying one particular conformation of the initial carbonyl compound in the rule is that in Grignard additions, RMgX is likely to coordinate not only with solvent ether but also with the carbonyl oxygen atom; the latter now almost certainly becomes the bulkiest group present in the molecule, and therefore tends to orient itself *between* the two least bulky groups attached to the adjacent dissymmetric carbon atom. In the cases considered above, establish-ment of the 'bulkiness sequence' as Ph > Me or Et > H presents no problem, but in less clear cut cases it can be a matter of some difficulty, with doubt then arising about the consequent prediction. Further, groups such as OH, OR, NH_2 etc. can themselves complex with, for example, a Grignard reagent and structures may then result which dictate the stereochemistry of the addition:

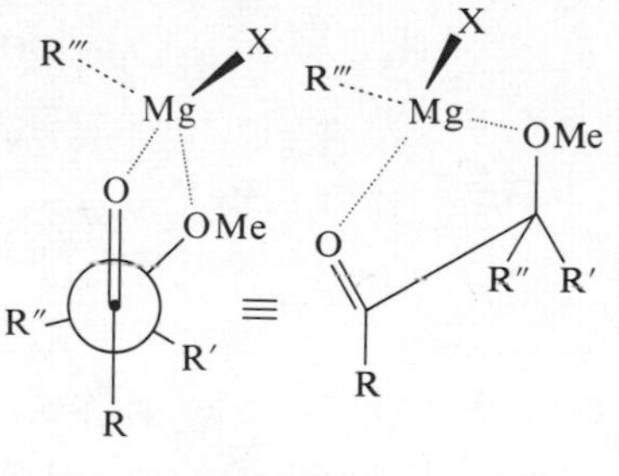

Thus with the carbonyl compound above, attack by $R'''MgX$ on the carbonyl carbon atom will take place preferentially from the least hindered side (from the right hand side if $R'' > R'$), but whether this will yield more of the diastereomer predicted by the Cram rule (which does not specifically allow for such an interaction) will depend on the relative sizes of OMe, R' and R''.

ELIMINATION REACTIONS

(a) Bimolecular (E2) polar eliminations

The observed products of bimolecular (E2) 1,2-elimination (in which the groups eliminated are originally situated on adjacent carbon atoms) from acyclic compounds are such that elimination must have taken place from a staggered conformation of the starting material in which the eliminated groups were in one plane and *trans* (*anti-*) to each other—the so-called *antiperiplanar* conformation; this clearly must be the condition of lowest energy demand. Thus the ($\pm$) stilbene dibromide (121) is found to undergo base-catalysed, TRANS dehydrobromination to yield the *trans*-bromostilbene (122), while the *meso*-dibromide (123) also undergoes TRANS dehydrobromination to yield the *cis*-bromostilbene (124):

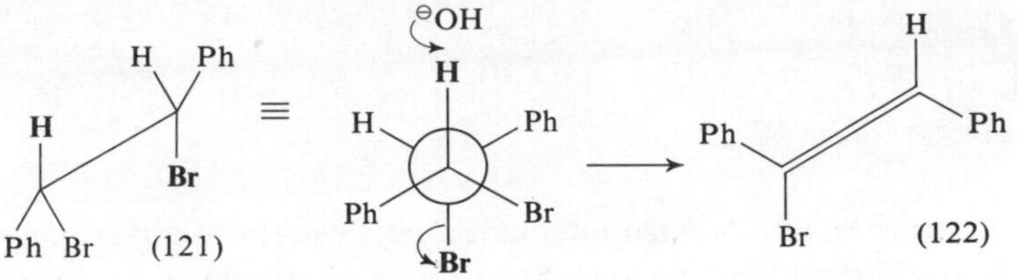

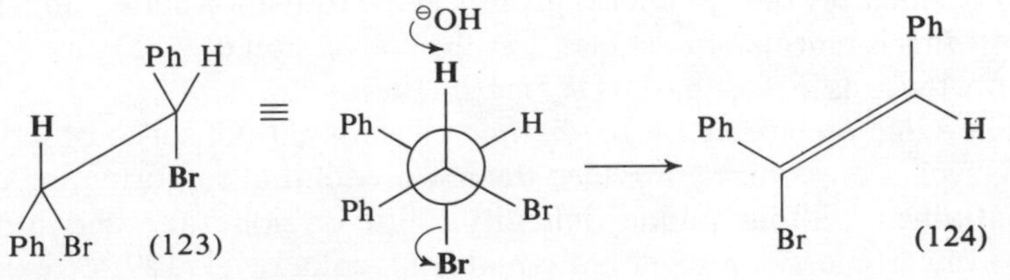

The reason for the preferred coplanarity is presumably to allow, in the transition state, the maximum overlap of the forming p orbitals on C_1 and C_2, thus ensuring the readiest—least energy-demanding—formation of the π bond in the product alkene. This requirement is apparently sufficiently important to overide the interaction that must result in the transition state from the close approach of the bulky phenyl groups in the conversion of (123) into (124). The effect of this phenyl/phenyl interaction is, however, reflected in the fact that, under comparable conditions, the conversion of (121) into (122) is very much more rapid than that of (123) into (124). This is sufficiently marked that on heating the corresponding dichlorides in pyridine at 200°, only the analogue of (121) undergoes elimination while that of (123) remains unchanged.

Elimination from the preferred *antiperiplanar* conformation is not confined to loss of hydrogen halide, the same is observed, for example, with loss of bromine from 1,2-dibromides (induced by iodide ion):

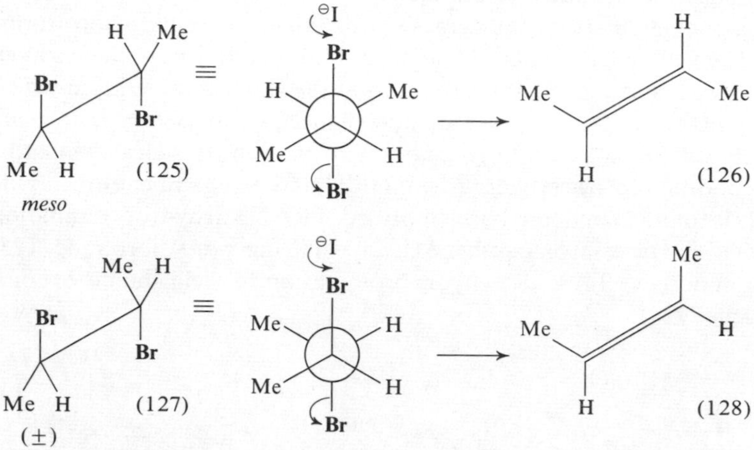

Here, however, non-bonded interaction between the methyl groups in the transition state leading to (128) will be much less marked than with the considerably bulkier phenyl groups in the transition state leading to (124); this is reflected in the fact that the conversion of (125) into (126) is only twice as fast as that of (127) into (128).

The same preference for *antiperiplanar* elimination is also observed with cyclic compounds, provided the required initial conformation can be attained without undue difficulty. With cyclohexane derivatives 1,2-*trans* isomers can be either equatorial–equatorial (129) or axial–axial (130), but it is only the latter that is *antiperiplanar* and from which the preferred TRANS elimination can thus take place:

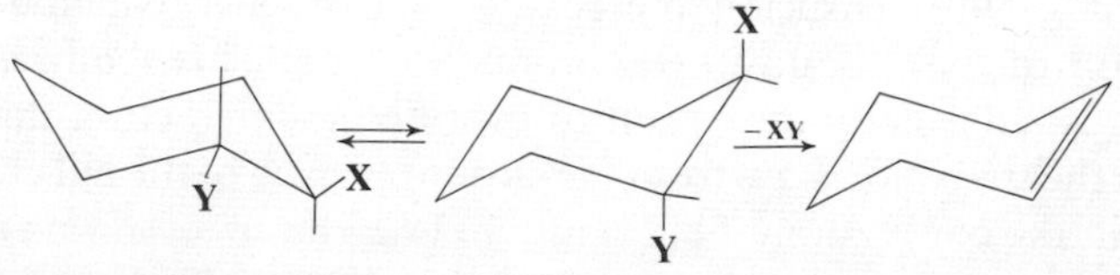

Normally (129) and (130) can be readily interconverted by 'flipping' of the ring as shown, and the desired axial–axial conformation thereby attained. The presence of a very bulky group elsewhere in the molecule may prevent this interconversion, however, for if it is sufficiently bulky it is precluded from assuming the axial state because of the non-bonded interaction with other axial substituents—even hydrogen—that would result. In the isomeric tosylates (131) and (132)—Ts = $p\text{-MeC}_6\text{H}_4\text{SO}_2$—the very bulky Me_3C group is thus restricted to the equatorial position; this imposes no limitation on (131) for the OTs and H on adjacent carbon atoms are both axial, i.e. ready elimination can take place from the *antiperiplanar* conformation:

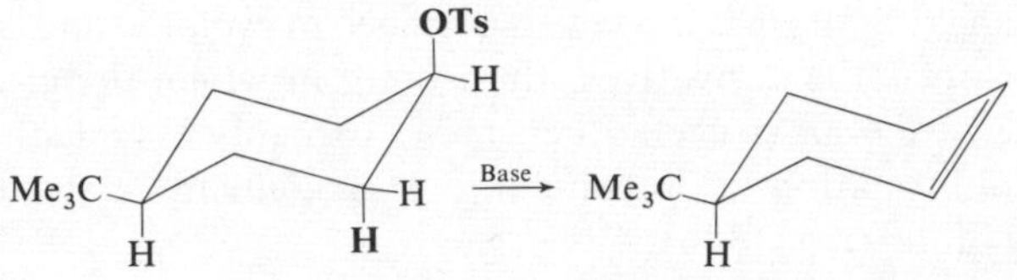

(131)

Though there is an axial H in the isomeric (132), the OTs is equatorial and cannot become axial by 'flipping' of the ring because the overriding need for the Me_3C group to be equatorial 'locks' the ring against such interconversion:

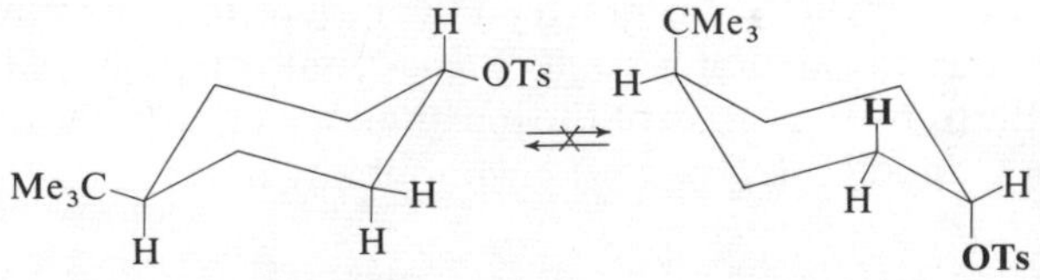

(132)

This 'locking' is reflected in the fact that (132) cannot be made to undergo bimolecular (E2) elimination at all; though some unimolecular elimination can be effected *via* a carbonium ion (E1), which has no critical stereochemical demand.

The preference for elimination from the 1,2-diaxial conformation is seen very clearly in the dehydrochlorination of the hexachlorocyclohexanes, $\text{C}_6\text{H}_6\text{Cl}_6$. There are a total of eight geometrical isomerides and of the five that have been investigated, four undergo elimination at very much the same rate, while the fifth reacts about 10,000 times more slowly. This latter isomer has been shown to have the structure (133), i.e. the one in which no H and Cl on adjacent carbon atoms are, or can become, *antiperiplanar*:

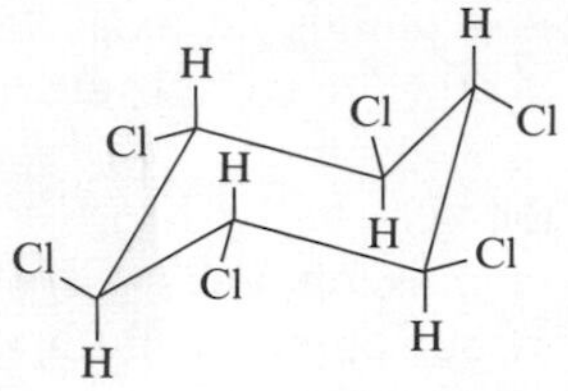

(133)

The possible attainment of the *antiperiplanar* conformation without introducing ring strain is, however, confined to cyclohexane derivatives. With cyclopentane derivatives, the isomer in which the groups to be eliminated are *trans* to each other, (134), can only assume the *antiperiplanar* conformation by introducing some element of strain through buckling of the ring (134b):

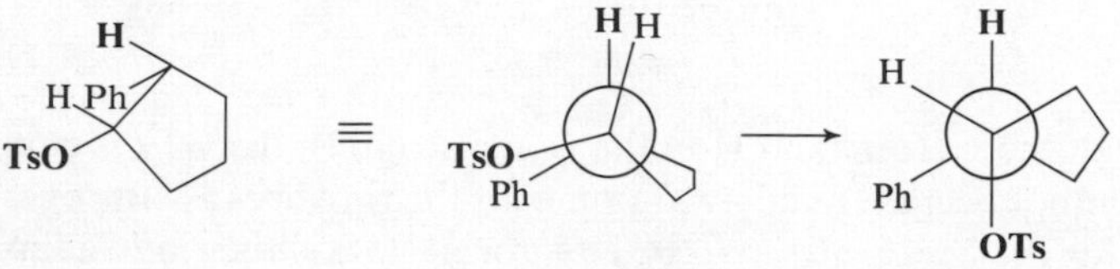

(134)   (134a)   (134b)

By contrast, the isomer (135) in which the groups to be eliminated are *cis* to each other can assume a *synperiplanar* conformation—which also allows, in the transition state, of the maximum overlap of the forming p orbitals on C_1 and C_2 (*cf.* p. 159)—without any distortion of the ring, (135a):

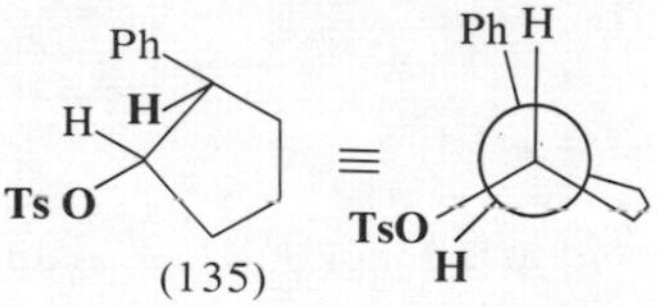

(135)

(135a)

Both these represent higher energy states—and by implication lead to higher energy transition states—than the 'ideal' *antiperiplanar* conformation: (134b) because of the ring strain introduced, (135a) because it represents a conformation in which the groups are eclipsed. We might therefore, with cyclopentane derivatives, expect less preference for TRANS elimination than with acyclic compounds or with cyclohexane derivatives, and it is found in fact that elimination from (134) with t-BuO$^\ominus$ is only about 10 times as fast as from (135): a very, very much smaller difference than is found with cyclohexane derivatives.

(b) Pyrolytic eliminations

There are a number of elimination reactions known, usually carried out by heating the substrate either alone or in an inert (non-solvating) solvent, that have a preferred, though not necessarily a 100 per cent, CIS stereochemistry. These are believed to proceed *via* cyclic transition states (from the *synperiplanar* conformation), though not necessarily without the intermediate development of some polar character. Thus the amine oxide (136) can act as its own intramolecular base, so far as proton removal is concerned, from the *synperiplanar* conformation (Cope reaction):

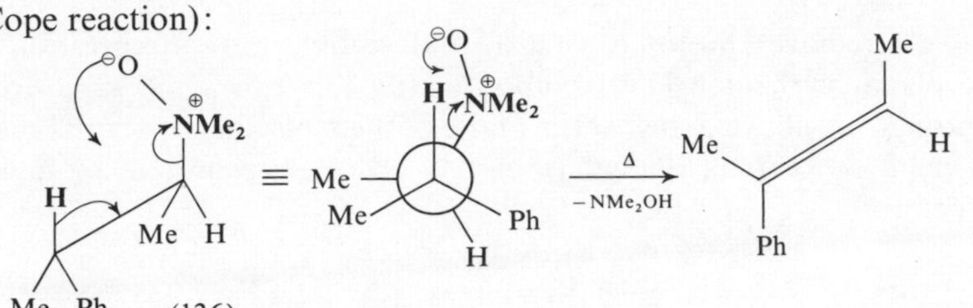

An analagous elimination occurs with xanthate esters such as (137),

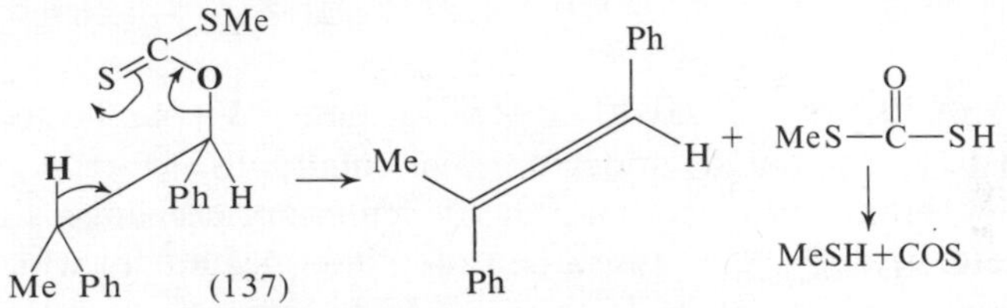

which is believed to proceed through a transition state of the form (138):

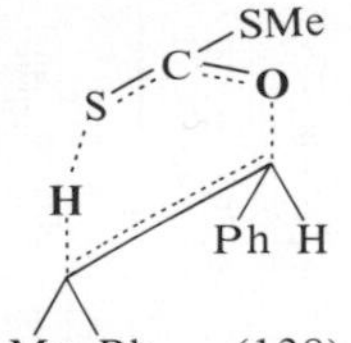

REARRANGEMENT REACTIONS

(a) Racemization in enolization etc.

When a dissymmetric centre, one of whose attached atoms is H, is adjacent to an electron-withdrawing group its optically active forms are found to undergo unusually ready racemization in the presence of base;

this is believed to involve the reversible formation of a carbanion, which must assume a planar state if the maximum stabilization by delocalization is to occur. Thus with the ketone (139),

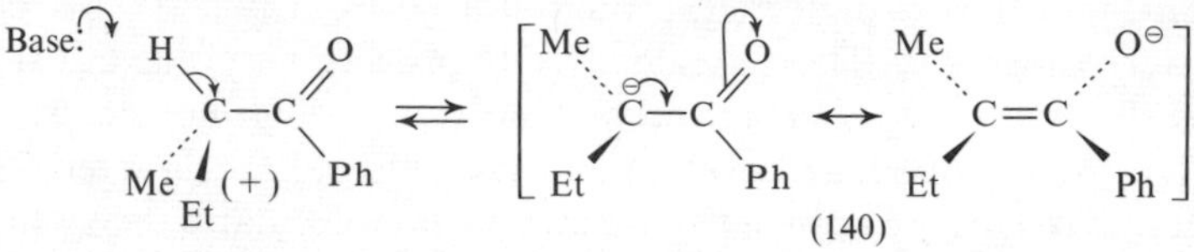

base can remove a proton to yield a small equilibrium concentration of the planar carbanion (140); subsequently this can again take up a proton, from above or below the plane of the molecule with equal ease, to yield the original optical isomer (139) or its enantiomer (139a), respectively:

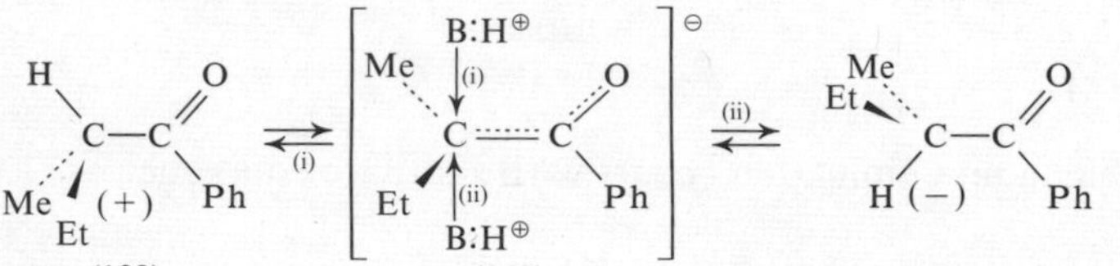

The plausibility of the carbanion intermediate (140) is reflected in the fact that (139) is found, under parallel conditions, to undergo racemization, bromination, iodination or deuterium exchange all at the same rate, strongly suggesting a common intermediate for which (140) is a prime candidate:

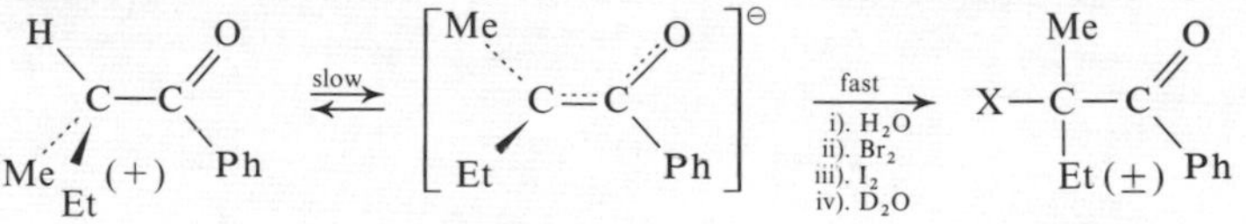

$$\text{(139)} \xrightarrow{\text{slow}} [\text{(140)}] \xrightarrow[\text{i). H}_2\text{O}~\text{ii). Br}_2~\text{iii). I}_2~\text{iv). D}_2\text{O}]{\text{fast}} X-\overset{Me}{\underset{Et\,(\pm)}{C}}-\overset{O}{C}\diagdown Ph$$

X = H, Br, I, or D, respectively

Similar ready racemization etc. of (139) is found to occur in the presence of acid, HA, though here the common intermediate is most probably the planar enol (141):

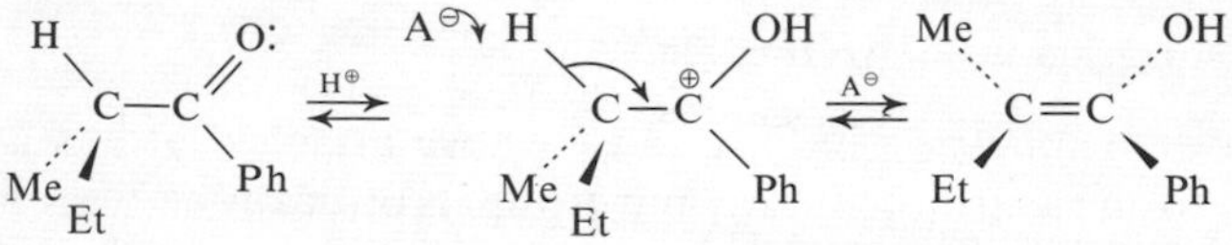

(b) Configuration of migrating groups

In rearrangements involving the migration of a group, Z, from one carbon, or other atom, to the adjacent one in a 1,2-shift (1,3-, 1,4-, etc. shifts are also known but are very much less common),

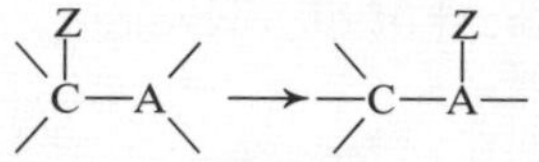

$$\begin{array}{ccc} \overset{Z}{\underset{|}{\diagdown}\!\!\!\diagup} & & \diagdown\;\overset{Z}{\underset{|}{\diagup}} \\ \diagup\!C\!-\!A\!\diagdown & \longrightarrow & \diagup\!C\!-\!A\!- \end{array}$$

it is a matter of some interest, if Z is dissymmetric, to determine whether its configuration is preserved or not during its migration. Where A is nitrogen, essentially total preservation has been demonstrated in a number of cases; thus the amide (142) is found, on undergoing the Hofmann reaction (*cf.* p. 75), to yield the amine (144) in which the C_6H_5MeCH group that has migrated from $C \longrightarrow N$ has wholly retained its original configuration:

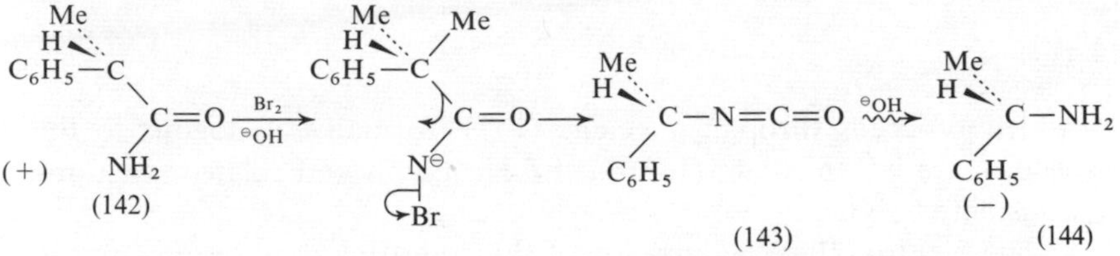

Similar preservations of configuration have also been observed in analogous $C \longrightarrow N$ shifts on other derivatives of the acid of which (144) is the amide, e.g. Lossen, Curtius and Schmidt reactions. Preservation of configuration also occurs in the $C \longrightarrow N$ shift in the Beckmann rearrangement of the ketoxime (145):

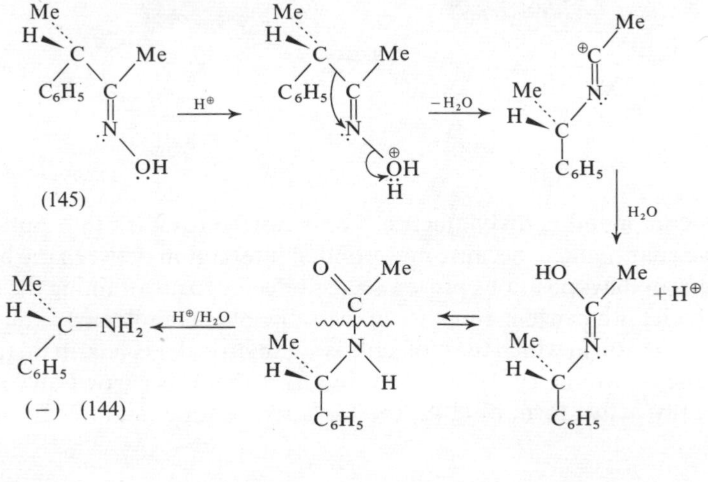

Preservation of configuration has been observed in a C —→ O shift, but direct, stereochemical observations on C —→ C shifts are very rare; this is surprising in view of the very large number of C —→ C shifts that have been investigated, albeit with other ends in view. One case where configuration has been shown to be preserved, however, is in the Wolff rearrangement of the diazoketone (146),

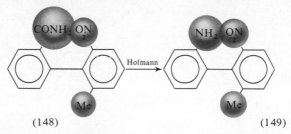

which proceeds through a ketene (147), formally analogous to the isocyanate intermediate (143) in the Hofmann and related reactions considered above.

The fact that the configuration of the migrating group is apparently always preserved constitutes powerful evidence that it never becomes to any significant extent free during its shift. There is a further ingenious piece of stereochemical evidence that directly demonstrates this to be the case; this involves the amide (148):

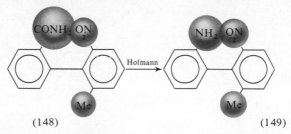

This compound is dissymmetric, i.e. it can be resolved into optically active enantiomers, because non-bonded interaction between the bulky groups in the *o*-positions prevents the molecule from attaining the state in which both benzene rings are in the same plane (coplanar)—the only conformation in which the molecule is symmetrical, i.e. has, in this case, a plane of symmetry. When a Hofmann reaction is carried out on an optically active form of (148) the biphenyl residue migrates from the carbon to the nitrogen atom of the original amide (*cf.* p. 75); if it became free in the process there would no longer be any restraint preventing its two benzene rings from becoming coplanar—one ring would now have no *o*-substituents—and the resultant amine (149) would thus be racemic (±). In fact, the amine that is obtained is still optically active, and the biphenyl residue thus did not become free during its migration.

The essentially universal preservation of configuration of the migrating group, and its consequent lack of freedom during migration, suggest a generalized reaction pathway of the form:

This would impose the further stereochemical requirements that the atom that Z migrates from, and the atom that it migrates to, should both undergo inversion of configuration: this too is indeed observed in the great majority of cases.

CONCLUSION

It will have been seen from the foregoing examples that stereochemical data provide important and detailed criteria by which the soundness of a projected reaction pathway may be judged; they may indeed be used in a number of cases to make the crucial, definitive choice between possible, alternative pathways. Stereochemical data are, however, seldom the best point from which to *begin* the study of a reaction mechanism: a good example of this is the S_N1 pathway of nucleophilic displacement where stereochemical data, (p. 128), as the initial criterion, would have been quite misleading. This is admirably expressed in the highly relevant dictum of Ingold: 'In general, stereochemistry without kinetic support is a poor tool for the investigation of mechanism: to start with it, is to start at the wrong end.'

FURTHER READING

BANTHORPE, D. V. 'The Transition States of Olefin-forming E2 Reactions', *Studies on Chemical Structure and Reactivity,* Ed. Ridd, J. H. (Wiley, 1966), pp. 131–145.

BOHM, B. A. and ABELL, P. I. 'Stereochemistry of Free Radical Addition to Olefins', *Chem. Rev.,* 1962, **62**, 599.

BORDWELL, F. G. 'Are Nucleophilic Bimolecular Concerted Reactions Involving Four or More Bonds a Myth?', *Acc. Chem. Res.,* 1970, **3**, 281.

168   *Stereochemical criteria*

BUNNETT, J. F. 'Olefin-forming Elimination Reactions', *Surv. Progr. Chem.* (Academic Press), 1969, **5**, 53.

BUNTON, C. A. 'Nucleophilic Substitution and the Walden Inversion', *Studies on Chemical Structure and Reactivity*, Ed. Ridd, J. H. (Wiley, 1966), pp. 73–102.

CAPON, B. 'Neighbouring Group Participation', *Quart. Rev.*, 1964, **18**, 45.

ELIEL, E. L. *Stereochemistry of Carbon Compounds* (McGraw Hill, 1962).

FAHEY, R. C. 'The Stereochemistry of Electrophilic Addition to Olefins and Acetylenes', *Topics in Stereochemistry* (Interscience), 1968, **3**, 237.

HÜCKEL, W. and HANACK, M. 'Eliminations in Cyclic *cis-/trans*-Isomers', *Angew. Chem. Int. Ed.*, 1967, **6**, 534.

MILLER, S. I. 'Stereoselection in the Elementary Steps of Organic Reactions', *Adv. Phys. Org. Chem.* (Academic Press), 1968, **6**, 185.

MISLOW, K. *Introduction to Stereochemistry* (Benjamin, 1966).

NEWAN, M. S. (Ed.) *Steric Effects in Organic Chemistry* (Wiley, 1956).

REUCROFT, J. and SAMMES, P. G. 'Stereoselective and Stereospecific Olefin Synthesis', *Quart. Rev.*, 1971, **25**, 135.

SKELL, P. S. 'Bridged Free Radicals', *Organic Reaction Mechanisms*, Special Pub. No. 19 (Chem. Soc., London), 1965, pp. 131–145.

TANIDA, H. 'Solvolysis Reactions of 7-Norbornenyl and Related Systems', *Acc. Chem. Res.*, 1968, **1**, 239.

TEDDER, J. M. 'The Interaction of Free Radicals with Saturated Aliphatic Compounds', *Quart. Rev.*, 1960, **14**, 336.

TRAYLOR, T. G. 'Electrophilic Addition to Strained Olefines', *Acc. Chem. Res.*, 1969, **2**, 152.

5
Structure/reactivity correlations

The rapidly increasing volume of data on equilibria and rates of organic reactions, coupled with the development of ideas about reaction mechanisms, has resulted in the development of a number of qualitative correlations between structure and reactivity, for example the rate sequence

$$CH_3CH_2Br > MeCH_2CH_2Br > Me_2CHCH_2Br \gg Me_3CCH_2Br$$

in the bimolecular S_N2 displacement reactions of the above halides (cf. p. 22). Clearly, the next step would be to establish, empirically, quantitative correlations in order (i) to clarify and categorize the existing data, (ii) to predict, reasonably accurately, the rate and equilibrium behaviour of compounds on which experimental data had not yet been obtained, and (iii) to provide general information about the mechanism of particular reactions.

HAMMETT PLOTS

The first such correlations were observed by Hammett (and also by Burkhardt) who found that plotting the log rate constants (log k) for various side chain reactions of m- and p-substituted benzene derivatives (e.g. the base initiated hydrolysis of substituted ethyl benzoates, m- or p-$XC_6H_4CO_2Et$) either against each other or against the log equilibrium constants (log K) for suitable reactions of similarly substituted benzene derivatives (e.g. the ionization of substituted phenyl-

169

acetic, *m*- or *p*-$XC_6H_4CH_2CO_2H$, or benzoic, *m*- or *p*-$XC_6H_4CO_2H$, acids in water at 25°) resulted in essentially straight lines:

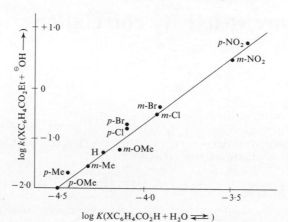

Fig. 5.1*-

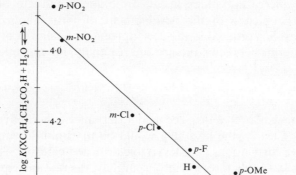

Fig. 5.2†

The straight line in Fig. 5.2 may be represented by an equation of the usual straight line form

$$\log K = \rho \log K' + c \qquad [1]$$

where ρ is the slope of the line and c the intercept. Designating the ionization constants of the unsubstituted parent acids, phenylacetic and benzoic acids (i.e. X=H), as K_o and K'_o, respectively, equation [1] becomes

$$\log K_o = \rho \log K'_o + c \qquad [2]$$

and then subtracting [2] from [1] yields

$$\log K - \log K_o = \rho(\log K' - \log K'_o)$$

or more concisely:

$$\log \frac{K}{K_o} = \rho \log \frac{K'}{K'_o} \qquad [3]$$

We can now take one of the two sets of equilibria as referring to a standard reaction: the ionization of m- and p-substituted benzoic acids in water at $25°$, K', was chosen because of the multiplicity and accuracy of experimentally determined K' values that were already available. We can then define a *substituent constant*, characteristic of a particular m- or a p-substituent X, by σ where $\sigma = \log K'_X/K'_o$, [3] thus becomes

$$\log \frac{K}{K_o} = \rho\sigma \qquad [4]$$

where ρ is a *reaction constant*, constant for any one reaction but varying from one reaction to another. σ will be constant for any given substituent, i.e. $\sigma_{p\text{-Me}}$ will differ from, for example, $\sigma_{m\text{-NO}_2}$ (and from $\sigma_{m\text{-Me}}$) but will itself be the same for the ionization of both p-$MeC_6H_4CH_2CO_2H$ and p-$MeC_6H_4CO_2H$ (these two reactions will, of course, have different ρ values), and for other reactions.

What we have done above, in designating a standard reaction, is to make ρ equal to $1·00$ for that standard reaction (the ionization of m- and p-substituted benzoic acids in water at $25°$); knowing, as we do, K values for a wide range of these substituted benzoic acids, K_o for benzoic acid itself, and ρ (by definition $= 1·00$), we can calculate $\sigma_{m\text{-}X}$ and $\sigma_{p\text{-}X}$ for a similarly wide range of substituents*:

Group	σ_{m-}	σ_{p-}	Group	σ_{m-}	σ_{p-}
t-Bu	$-0·10$	$-0·197$	Cl	$+0·373$	$+0·227$
Me	$-0·069$	$-0·170$	COMe	$+0·376$	$+0·502$
H	$0·000$	$0·000$	Br	$+0·391$	$+0·232$
OMe	$+0·115$	$-0·268$	CN	$+0·56$	$+0·660$
OH	$+0·121$	$-0·37$	NO$_2$	$+0·710$	$+0·778$
F	$+0·337$	$+0·062$			

*Reproduced from D. H. McDaniel and H. C. Brown, *J. Org. Chem.*, 1958, **23**, 420 by permission of the copyright (1958) holder, the American Chemical Society.

Using these substituent constants and the experimentally determined K (and K_o) values for another reaction, e.g. the ionization of phenylacetic acids, $ArCH_2CO_2H$, it is now possible to determine the value of ρ for this reaction too; it turns out to be 0·471 (*cf.* 6.) below). Values for ρ for a wide variety of other reactions† may be obtained in a similar manner:

	Reaction	Type	ρ
1.	$ArNH_2$ with $2,4\text{-}(NO_2)_2C_6H_3Cl$ in EtOH at 25°	k	−3·190
2.	$ArNH_2$ with $PhCOCl$ in benzene at 25°	k	−2·694
3.	$ArCH_2Cl$ hydrolysis in aq. acetone at 69·8°	k	−1·875
4.	$ArO^{\ominus}$ with EtI in EtOH at 25°	k	−0·991
5.	$ArCO_2H$ with MeOH (acid-catalysed) at 25°	k	−0·085
6.	$ArCH_2CO_2H$ ionization in H_2O at 25°	K	+0·471
7.	$ArCH_2Cl$ with $I^{\ominus}$ in acetone at 20°	k	+0·785
8.	$ArCH_2CO_2Et$ hydrolysis (base) in aq. EtOH at 30°	k	+0·824
9.	$ArCO_2H$ ionization in H_2O at 25°	K	+1·000 (standard)
10.	$ArOH$ ionization in H_2O at 25°	K	+2·008
11.	$ArCN$ with H_2S in alkaline EtOH at 60·6°	k	+2.142
12.	$ArCO_2Et$ hydrolysis (base) in aq. EtOH at 30°	k	+2·498
13.	$ArNH_3^{\oplus}$ ionization in H_2O at 25°	K	+2·730

Log K values are related to standard free energy changes by the equation (*cf.* p. 4)

$$\Delta G^\circ = -2\cdot303 \, RT \log K$$

$$\text{i.e.} \quad \log K = \frac{-\Delta G^\circ}{2\cdot303 \, RT} \qquad [5]$$

and the straight line relationships empirically established, at constant temperatures, between sets of log Ks for two different equilibria, e.g. the ionization of m- and p-substituted-benzoic and -phenylacetic acids in water at 25° (*cf.* Fig. 5.2), thus relate to ΔG° differences; these approximately straight line plots are generally referred to as *linear free energy relationships*.

While it is thus not perhaps unreasonable that sets of log Ks for two different reactions should be so related, the observed straight line relationship between a set of log Ks for one reaction and a set of log ks (rate constants) for another (*cf.* Fig. 5.1) seems rather more surprising; at least until we remember that log k can be related to $\Delta G^{\ddagger}$ (the free energy of activation) by an equation of the same general form as [5]. We then realize that here, too, free energy change differences of a similar character are involved for log k, which determines $\Delta G^{\ddagger}$, can,

---

†Reproduced from L. P. Hammett, *J. Amer. Chem. Soc.*, 1937, **59**, 97 by permission of the copyright (1937) holder, the American Chemical Society.

in terms of transition state theory, be considered as related to an equilibrium constant for the reaction converting starting materials into the transition state: a linear free energy relationship again results.

The simplest use of these relationships would, as mentioned above, be to predict rate or equilibrium behaviour for a compound or compounds on which experimental data have not yet been obtained. Thus in the base-initiated hydrolysis of ethyl *m*- and *p*-substituted benzoates (1)

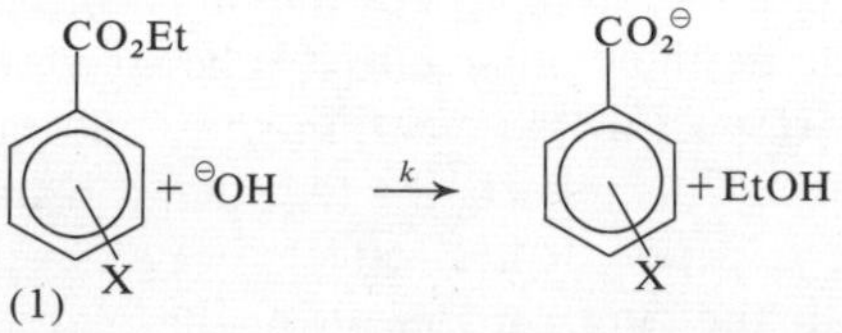

(1)

it was found that the *m*-nitroester (X = m-NO_2) was hydrolysed 63·5 times as fast as ethyl benzoate itself (X = H); $\sigma_{m\text{-}NO_2}$ has a value (from the ionization of substituted benzoic acids) of 0·710 (see above), and substitution in equation [4] (p. 171), thus enables us to evaluate ρ for the reaction:

$$\log \frac{63\cdot5}{1} = \rho \times 0\cdot710$$

$$\rho = 2\cdot54$$

We are now able to calculate (using equation [4]) the rate of hydrolysis, relative to the rate of hydrolysis of ethyl benzoate, for any ethyl *m*- or *p*-substituted benzoate for whose substituent we know the value of σ. Thus $\sigma_{p\text{-}OMe}$ has a value of $-0\cdot268$ (*cf.* p. 171) and hence

$$\log \frac{k_{p\text{-}OMe}}{k_H} = 2.54 \times -0\cdot268$$

so that: $\quad \dfrac{k_{p\text{-}OMe}}{k_H} = 0\cdot209$

i.e. ethyl *p*-methoxybenzoate is predicted to hydrolyse 0·209 times as fast as ethyl benzoate itself: the experimentally determined value was found to be 0·214.

Other, somewhat more sophisticated, uses of Hammett plots are discussed below.

THE PHYSICAL SIGNIFICANCE OF σ AND ρ

Having established the empirical validity of the parameters σ and ρ, it now remains to attempt to give them some justification in terms of the familiar factors that influence rates and equilibria. Scrutiny of the $\sigma_{m\text{-}}$

values for substituents in the *m*-position—from whence they can exert only an inductive effect through the bonds, and a possible field effect through the medium, on the reaction centre—suggests that $\sigma_{m\text{-}}$ may be a measure of an overall polar effect exerted by the substituent. The $\sigma_{m\text{-}}$ values are seen to be $-$ve for an electron-donating (compared with H) substituent, $+$ve for an electron-withdrawing one, their magnitude depending on the size of the effect. The $\sigma_{p\text{-}}$ values are found not only to differ in magnitude from $\sigma_{m\text{-}}$ for the same substituent (they would be expected to do as their polar effect is being exerted from a different position with respect to the reaction site), but may also differ in sign (e.g. OMe, OH). This reflects the fact that a *p*-substituent can, in addition to any inductive or field effects it may exert, also interact conjugatively with the ring carbon atom that carries the side chain at which reaction is taking place, e.g. in the base-initiated hydrolysis of ethyl *m*- and *p*-methoxybenzoates (2 and 3, respectively):

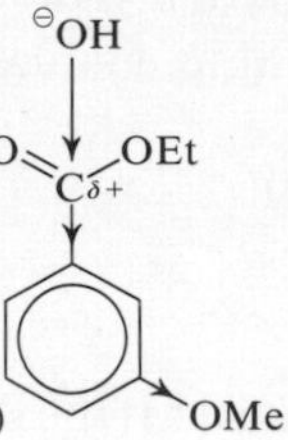

$$\sigma_{m\text{-OMe}} = +0\cdot 115$$

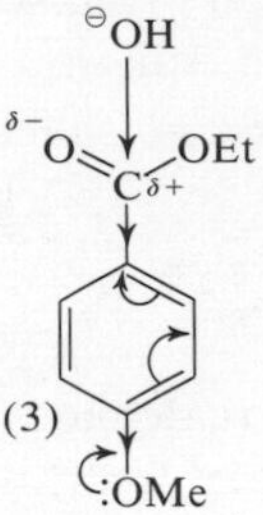

$$\sigma_{p\text{-OMe}} = -0\cdot 268$$

Here, *m*-OMe is electron-withdrawing in (2), as no doubt, inductively, *p*-OMe is in (3), but the electron-donating conjugative effect that OMe can exert from the *p*- (though not from the *m*-) position sufficiently outweighs the inductive effect to make *p*-OMe, overall, electron-donating. Where direct conjugative interaction can take place between a *p*-substituent and the reaction centre itself (5), for example in the alkylation of aryloxide ions (4 and 5)

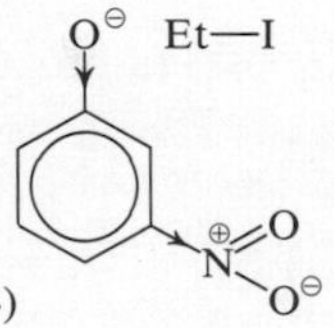

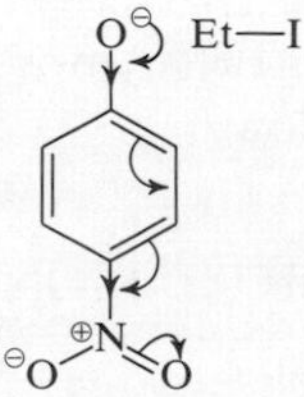

the usual $\sigma_{p\text{-}}$ values for such substituents no longer fall on a straight line as we shall see below (p. 183).

Just as σ_X is a measure of the polar response that a substituent X is able to make, ρ, the reaction constant, is a measure of a reaction's polar requirement compared to that of the standard reaction (the ionization of $ArCO_2H$ in water at 25°): polar requirement essentially meaning susceptibility, compared with the standard reaction, to electron-donation/-withdrawal. It will thus follow that:

1. Reactions that are facilitated by electron-withdrawal will have +ve values for ρ, and *vice versa*.
2. Reactions that are more susceptible to polar effects than the standard reaction will have ρ values (+ve or −ve) greater than unity (ρ for the standard reaction was defined as 1·00).

The ρ values for the ionization of phenols, ArOH, and anilinium ions, $ArNH_3^{\oplus}$, are +ve like that for the ionization of benzoic acids, $ArCO_2H$—all are assisted by electron-withdrawing groups—but their values are larger than 1·00 (the value for benzoic acids) because the polar effects of the substituents are operating on an atom (O and N, respectively) immediately adjacent to the benzene nucleus, whereas in the benzoic acids these effects have to operate on an atom (O∿H) one further removed. Similarly, ρ for the ionization of the phenylacetic acids, $ArCH_2CO_2H$, while still +ve is smaller than 1·000, because the polar effect of substituents has to operate on an atom one further removed again from the benzene ring. By contrast, the benzoylation of aromatic amines, $Ar\dot{N}H_2$, will clearly be assisted by electron donating groups, and ρ is indeed seen to have a fairly large −ve value.

USE OF HAMMETT PLOTS

A simple example of the use of Hammett plots to predict a reaction rate has been described above (p. 173). The major use of such plots lies, however, not in making simple predictions of this nature, but in establishing σ values for new substituents and, especially, ρ values for new reactions. These values can then be used, in more general terms, to provide information about the mechanism of particular reactions; this is especially true of the magnitude and sign of ρ values. Thus a +ve value for ρ indicates the development of a partial −ve charge at the reaction centre during the formation of the transition state (and/or intermediate), and *vice versa*. The magnitude of the ρ value is thus a measure of the magnitude of the developing charge and of its interaction with the substituent. Hardly surprisingly, these mechanistic insights are somewhat less reliable when the magnitude of the ρ value approaches zero.

One implied, though not as yet specifically stated, requirement for a linear free energy relationship is that there should be no change of a particular reaction's mechanism as the substituents are varied; if there is such a change, then a straight line plot is no longer obtained. It thus follows that actually obtaining a non-linear plot may be a useful indication of change in mechanism, or of change in the rate-limiting step within the same, overall mechanistic scheme. A good example of this is seen in the reaction of *m*- and *p*-substituted benzaldehydes (6) with semicarbazide to form semicarbazones (7):

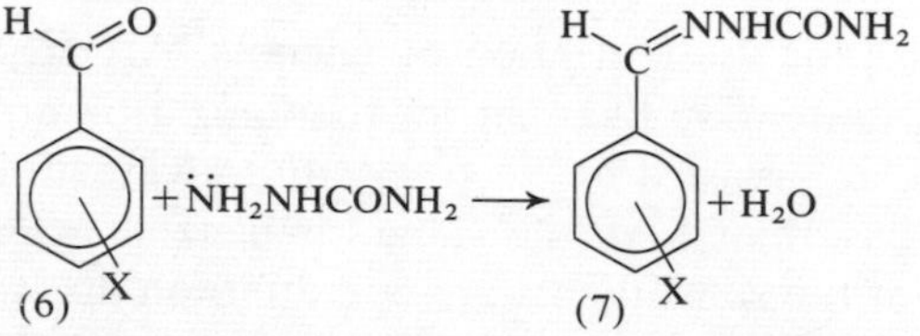

The rate of this reaction was found to vary with pH: on going from slightly basic to slightly acidic conditions the overall rate was found to increase, but then to decrease again as the conditions were made more strongly acidic. In order to account for this behaviour a scheme of the following form (*cf.* p. 82) was suggested

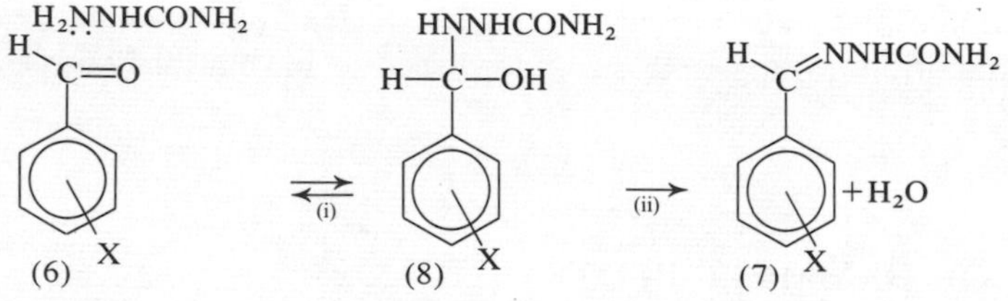

in which nucleophilic addition to (6) is followed by acid-catalysed dehydration of the initial adduct (8) to yield the final product (7). At slightly alkaline pH (ii) will be the slow, rate-limiting step, but as the pH is made slightly more acidic this acid-catalysed step will be speeded up, and the overall rate will increase. As the pH is made more acidic this step will continue to speed up but the nucleophilic semicarbazide will be increasingly protonated,

$$AH + :NH_2NHCONH_2 \rightleftharpoons A^\ominus + H:\overset{\oplus}{N}H_2NHCONH_2$$

and thus less able to act as a nucleophile; step (i) will thus slow down until it becomes the rate-limiting step, and the overall reaction rate (k observed) will become progressively slower.

An excellent test of the validity of this reaction pathway, and the variation of its rate-limiting step with pH, is provided by Hammett plots for a range of *m*- and *p*-substituted benzaldehydes at different pHs. Thus at pH 1·75 (relatively highly acidic) the plot of σ against log $k_{obs.}$ for the reaction of the series of ArCHOs with semicarbazide in 25 per cent ethanol at 25° is represented by Fig. 5.3:

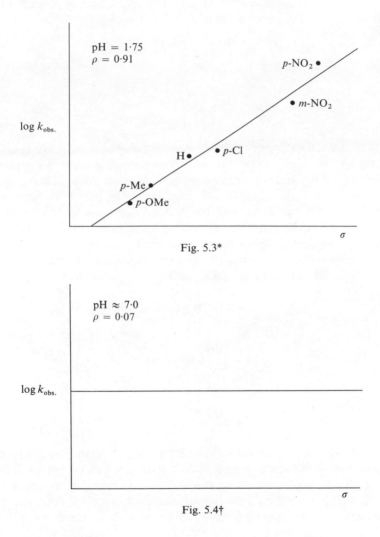

Fig. 5.3*

Fig. 5.4†

*Reproduced from B. M. Anderson and L. P. Jencks, *J. Amer. Chem. Soc.*, 1960, **82**, 1775 by permission of the copyright (1960) holder, the American Chemical Society.

†Reproduced from B. M. Anderson and L. P. Jencks, *J. Amer. Chem. Soc.*, 1960, **82**, 1774 by permission of the copyright (1960) holder, the American Chemical Society.

The slope, $\rho = 0.91$, indicates that the rate-limiting step of the reaction under these conditions is facilitated by electron-withdrawing substituents (fastest with X=p-NO$_2$, slowest with X=p-OMe). This is in conformity with step (i)—attack by the nucleophile—being rate-limiting, for electron-withdrawing-substituents will make the carbonyl carbon atom in ArCHO more positive (compared with PhCHO: 9, i.e. X=H) and thus more readily attacked by the electron pair on the nitrogen atom of semicarbazide, e.g. when X=m-NO$_2$ (10):

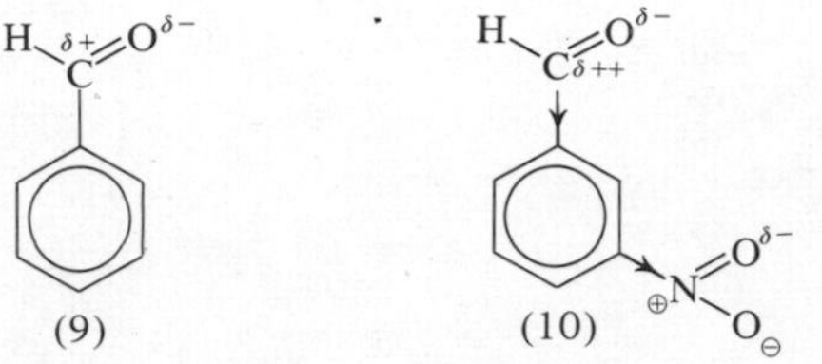

(9)   (10)

By contrast the slope of the corresponding plot at $\approx$ pH 7·0 is found to be $\rho = 0.07$; that is to say the overall reaction rate is now essentially uninfluenced by electron-withdrawal or -donation. Step (i) must still be in operation, and the observed slope, or rather lack of it, suggests therefore that a step facilitated by electron-donation must also now be involved in influencing the overall reaction rate. This must be the dehydration which involves the development of some +ve character on the carbonyl carbon atom in the transition state (12), and so will be facilitated by electron-donating substituents, e.g. when X = p-Me (11):

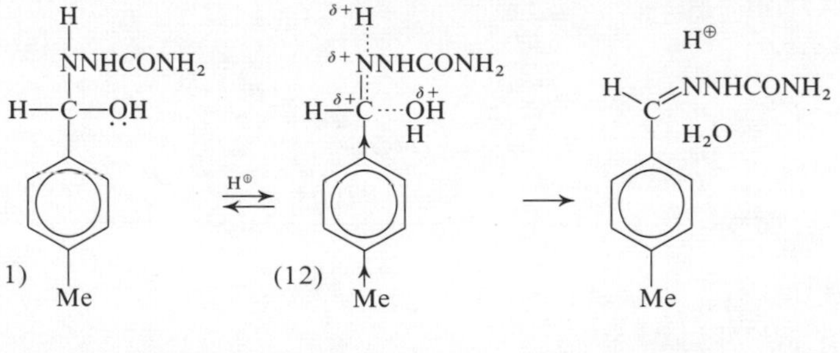

(11)   (12)

Step (ii)—the acid-catalysed dehydration—is indeed the rate-limiting step of the overall reaction under these conditions and it may well be asked why the non rate-limiting step (i) is still exerting an influence, as exemplified by its effect on the overall electronic needs of the reaction. The answer is that step (i) is an equilibrium and the larger the value of its K, the greater will be the concentration of the initial adduct (8), and the higher the concentration of adduct the more rapidly will final product, the semicarbazone (7), be produced from starting material.

Electron-withdrawing substituents influence K for step (i) in essentially the same way that they influence k for its formation (see above), i.e. the more electron-withdrawing the substituent the larger the value of K. Hence the rate of final product formation (i.e. overall reaction rate) will be facilitated by both electron withdrawing (larger K in step (i)) and electron-donating (larger k in step (ii)) substituents; these effects essentially cancel out and the result is a plot with a ρ value of only 0·07.

The acid test of the above explanations would be the observation at an intermediate pH, of a $\log k_{obs.}/\sigma$ plot in which aldehydes with electron-donating substituents fall on one slope and those with electron-withdrawing substituents on another, the change hinging on X=H, i.e. a change of rate-limiting step, at constant pH, depending on the substituent: at pH 3·9, this is just what is observed (Fig. 5.5):

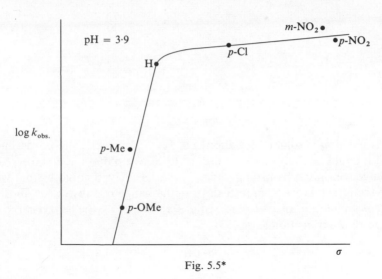

Fig. 5.5*

At this pH, the rate-limiting step is attack of the nucleophile to form the adduct (8) for aldehydes with electron-donating substituents (X=p-OMe, p-Me), the parent aldehyde (X=H) is just about at the borderline, and the rate-limiting step is dehydration of the adduct (8) to yield product (7) for aldehydes with electron-withdrawing substituents (X=p-Cl, m-NO$_2$, and p-NO$_2$). The three plots (Figs. 5.3, 5.4 and 5.5) thus neatly codify the experimental data, thereby providing effective support for the suggested reaction scheme.

*Reproduced from B. M. Anderson and L. P. Jencks, *J. Amer. Chem. Soc.*, 1960, **82**, 1775 by permission of the copyright (1960) holder, the American Chemical Society.

It should be remarked that as sharp a transition from one straight-line plot to another as that in Fig. 5.5 is relatively uncommon, more usual is a relatively smooth curve as in the reaction of $ArCH_2Cl$ with trimethylamine in benzene at 100° (Fig. 5.6; k/k_o is the relative rate with respect to that of $C_6H_5CH_2Cl$ itself):

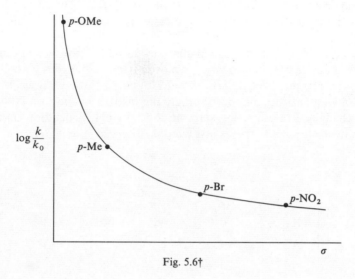

Fig. 5.6†

This reaction is somewhere about an S_N1/S_N2 mechanistic borderline and it could be that we are seeing in the above plot an actual discrete changeover in mechanistic pathway. Alternatively, it could be that we are seeing a relatively smooth shift, with change of substituent, in the extent and importance of generating +ve charge on the benzyl carbon atom in the transition state (13)

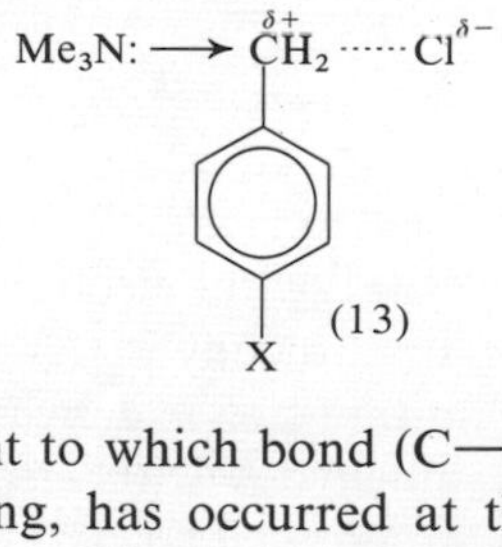

$$Me_3N: \longrightarrow \overset{\delta+}{C}H_2 \cdots Cl^{\delta-}$$

(13)

i.e. of the relative extent to which bond (C—Cl) breaking, compared to bond (N—C) making, has occurred at the transition state. The former will clearly be assisted by electron-donating substituents such as p-OMe and p-Me. It is significant in this respect that the plot

†Reproduced from C. G. Swain and W. P. Langsdorf Jr., *J. Amer. Chem. Soc.*, 1951, **73**, 2815 by permission of the copyright (1951) holder, the American Chemical Society.

for *m*-substituted derivatives—in which conjugative electron-donation by substituents cannot occur—while not a straight line is a very much shallower curve than the above.

A recent example is in the acetolysis of 3-aryl-2-butyl brosylates

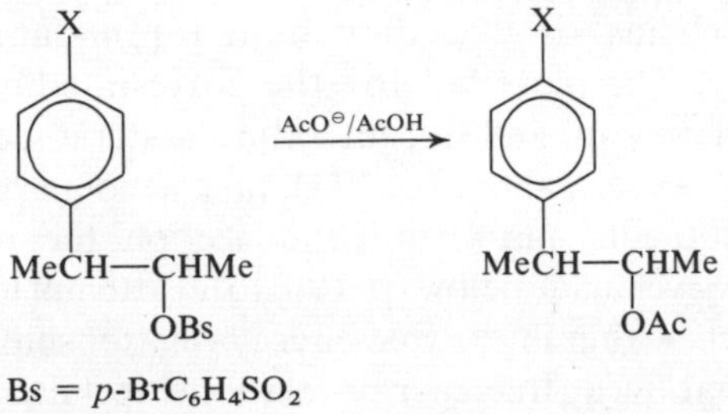

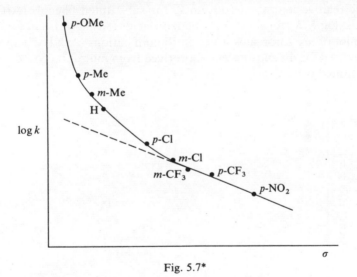

$$Bs = p\text{-}BrC_6H_4SO_2$$

for which a Hammett plot of the total observed rate constant against σ is found to have the form:

The plot is thus a straight line for electron-withdrawing substituents but curves upwards for electron-donating substituents; this has been interpreted as demonstrating simple solvolysis operating in the straight line portion, but with increasing aryl participation—as a neighbouring group (*cf.* p. 133)—as X becomes more electron-donating.

*Reproduced from H. C. Brown, P. v. R. Schleyer *et al.*, *J. Amer. Chem. Soc.*, 1970 **92**, 5244 by permission of the copyright (1970) holder, the American Chemical Society.

LIMITATIONS OF, AND DEVIATIONS FROM, HAMMETT PLOTS

Perhaps the most obvious limitation of Hammett plots—quite apart from the over-riding limitation of holding only for the reactions of aromatic side chains—is that they hold for *m*- and *p*-substituted derivatives only. The plots fail for the corresponding *o*-substituted derivatives as these can, and do, introduce a steric factor (*cf.* p. 116) whereas σ and ρ, as seen above (p. 173), take account only of the polar effects of substituents and their influence on the reaction centre. Reference is however made below (p. 190) to an attempt to accommodate such steric effects within linear free energy relationships.

It is found that linear free energy relationships also break down in part for some reactions of *p*-substituted derivatives. Thus Fig. 5.8 is a plot of log K/K_0 (i.e. ionization constant for the compound with substituent X compared with that for the compound where X=H, hence relative ionization constants) for substituted phenols (ArOH) against log K/K_0 for substituted benzoic acids (ArCO$_2$H), and Fig. 5.9 is a plot of log K for substituted anilinium cations (ArNH$_3^{\oplus}$) against values of σ for the substitutent X (derived from ionization constants of substituted benzoic acids, *cf.* p. 171);

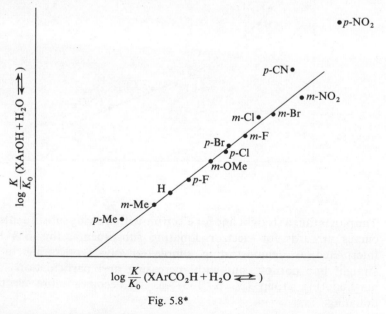

$$\log \frac{K}{K_0} (\text{XArCO}_2\text{H} + \text{H}_2\text{O} \rightleftharpoons)$$

Fig. 5.8*

*Reproduced from R. W. Taft Jr. and I. C. Lewis, *J. Amer. Chem. Soc.*, 1958, **80**, 2437 by permission of the copyright (1958) holder, the American Chemical Society.

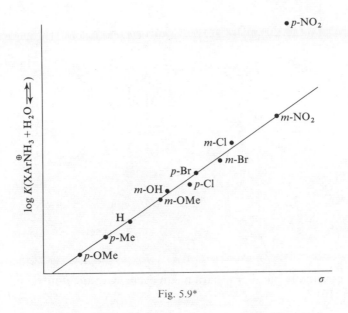

Fig. 5.9*

It will readily be seen that in both cases the p-NO_2 substituent (and, to a lesser extent, the p-CN substituent in Fig. 5.8) lies well off the line: both p-nitrophenol (14) and the p-nitroanilinium cation (16) are stronger acids than would have been predicted from the known effect of the p-NO_2 substituent on the acidity of benzoic acid. Clearly, in these cases, the p-NO_2 group is able to exert an additional acid-strengthening effect that it cannot exert in benzoic acid; this arises because the reaction site, with both phenols and anilinium cations, is the atom adjacent to the benzene nucleus, and conjugative interaction with the ring, and hence with its electron-withdrawing p-NO_2 substituent, can thus take place. With the phenol (14) such conjugative interaction can take place in the unionized starting material (14 ←→ 14a), and also in the anion (15 ←→ 15a). The resultant delocalization will serve to stabilize both, but whereas in (14 ←→ 14a) charge-separation is involved, for which an energetic price would be exacted, with (15 ←→ 15a) it is not; the stabilization is thus considerably more marked in the anion than in the phenol itself, ionization is thus promoted, and the phenol is a stronger acid than we would have expected from the effect of p-NO_2 on benzoic acid, where such conjugative interaction with the reaction centre cannot occur.

*Reproduced from J. Hine, 'Physical Organic Chemistry' (2nd Ed., 1962), p. 89, by permission of the copyright (1962) holder, McGraw Hill Book Co.

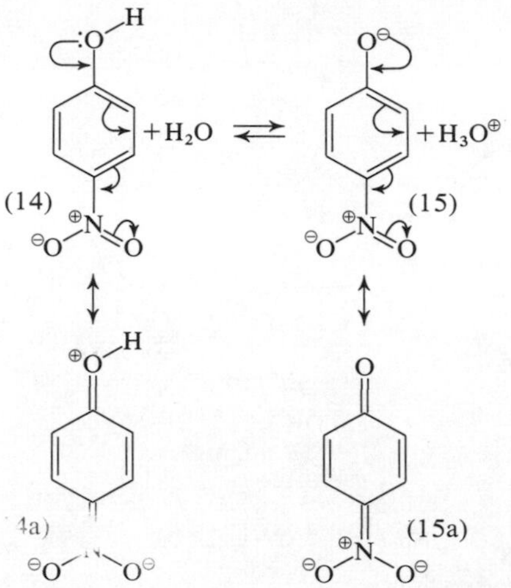

(14) +H₂O ⇌ (15) +H₃O⊕

(14a)   (15a)

An essentially parallel situation can occur with the *p*-nitroanilinium cation (16):

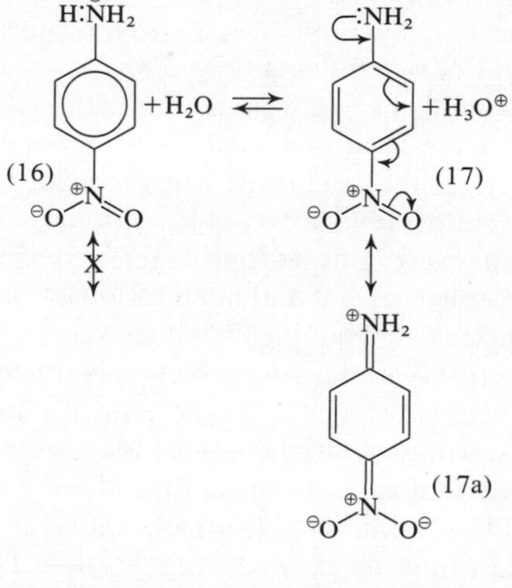

(16) +H₂O ⇌ (17) +H₃O⊕

(17a)

Here, conjugative interaction in the conjugate base (17 ⟷ 17a) involves charge separation and will, to that extent, be the less effective as a stabilizing influence, but analagous conjugative interaction cannot, by contrast, take place at all in the cation (16) as the electron pair on nitrogen is not available, being bonded to a proton. (17 ⟷ 17a) is thus highly stabilized with respect to (16), whose ionization is thus considerably promoted.

These conjugative influences are, of course, polar effects, but not ones that can operate either from the *m*-position, or in the ionization of benzoic acids (*cf.* p. 183) from which our substituent σ values were obtained. For reactions in which such conjugative effects can operate we shall therefore need alternative σ_{p-} values, for at least some substituents, to take account of them: thus while the normal σ value for *p*-NO$_2$ is $+0.778$, we find we require another, different value of $+1.27$ to account for the effect of this substituent on the ionization of phenols, or of anilinium cations, and for some other reactions.

The above examples are deviations from linearity arising from conjugative interaction between an electron-withdrawing *p*-substituent and the increased electron density developing at the reaction centre: the atom adjacent to the nucleus. We might also expect deviations to arise from conjugative interaction between an electron-donating *p*-substituent and any decreased electron density (i.e. positive charge) developing at the reaction centre in the transition state, if this too involves the atom next to the nucleus: this is indeed observed. Thus the solvolysis of substituted phenyldimethylcarbinyl (cumyl) chlorides (18) in 90 per cent aqueous acetone is essentially S$_N$1

$$\text{Me}_2\overset{|}{\text{C}}\text{—Cl} \quad\quad \text{Me}_2\text{C}^{\oplus}\text{Cl}^{\ominus} \quad\quad \text{Me}_2\overset{|}{\text{C}}\text{—OH}$$

$$\underset{(18)\ X}{\text{⬡}} \quad \xrightarrow{\text{slow}} \quad \underset{(19)\ X}{\text{⬡}} \quad \xrightarrow[\text{fast}]{\text{H}_2\text{O}} \quad \underset{X}{\text{⬡}}$$

the carbonium ion pair (19) being an intermediate. An electron-donating *p*-substituent, e.g. OMe, can stabilize such an intermediate by conjugative delocalization (20 ⟷ 20a)

$$\text{Me}\diagdown\overset{\oplus}{\underset{|}{\text{C}}}\diagup\text{Me} \quad\quad\quad \text{Me}\diagdown\underset{||}{\text{C}}\diagup\text{Me}$$

$$\underset{(20)\quad \overset{..}{\text{C}}\text{:OMe}}{\text{⬡}} \quad \longleftrightarrow \quad \underset{\overset{\oplus}{\text{OMe}}\quad (20a)}{\text{⬡}}$$

and, in so far as (20) is a model for the preceding transition state, the latter will also be stabilized, i.e. its energy level will be lowered, and it will thus be attained more readily: the reaction rate will therefore be increased. It is found in fact that *p*-OMe is well off the straight line for the plot of log k_X/k_H against σ (values derived from ArCO$_2$H ionizations), $k_{p\text{-OMe}}$ being about 100 times larger than predicted. Thus for this type of side chain reaction too another, different, value of σ will be required for substituents such as *p*-OMe. We shall return to this question of dual σ values below (p. 188).

AROMATIC ELECTROPHILIC SUBSTITUTION

An obvious possible extension of the field of Hammett plots would be to electrophilic substitution, e.g. nitration, sulphonation, halogenation etc., *m*- and/or *p*- to a substituent already present in the benzene nucleus:

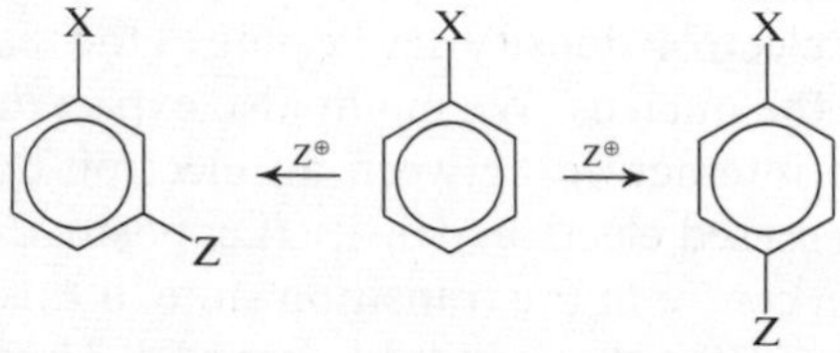

Attack on the *o*-position would again be neglected because of the steric effects that are known to be involved. The intention would be to plot the known substituents σ values, obtained from $ArCO_2H$ ionizations, against the log of the measured relative rates, log k/k_0, for a particular reaction to see if a straight line—whose slope defines ρ for the reaction—is obtained. There is a slight complication over the rate measurements in that k_0 refers to benzene, the standard, which has *six* replaceable hydrogens (21), while k refers either to attack on *two* alternative *m*-positions (22) in C_6H_5X, i.e. k_{m-}, or to attack on only *one* *p*-position (23), i.e. k_{p-}: a statistical factor has thus to be introduced:

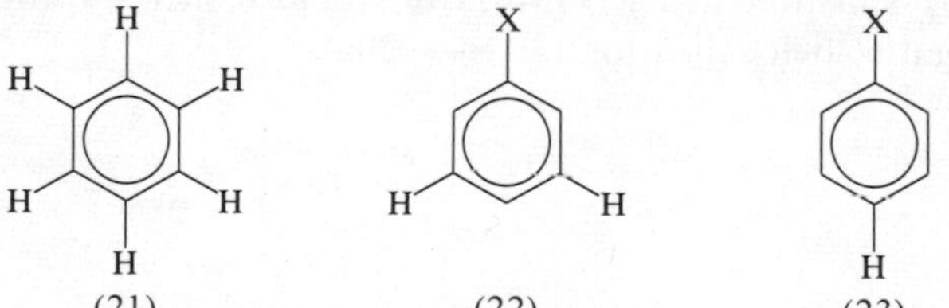

(21)   (22)   (23)

Actual experimental data normally obtained comprise an overall rate for benzene ($k_{C_6H_6}$), an overall rate for C_6H_5X ($k_{C_6H_5X}$)—or a ratio of the two ($k_{C_6H_5X}/k_{C_6H_6}$), and an analysis (by spectroscopic or other means) of the relative proportions of *o*-, *m*- and *p*-isomers obtained (generally expressed as percentages of the total—summed—monosubstitution products). We can then evaluate the k_m/k_H or k_p/k_H that we need by:

$$\frac{k_{m-}}{k_H} = \frac{k_{C_6H_5X}}{k_{C_6H_6}/\underline{6}} \times \frac{\%m\text{-product}}{\underline{2} \times 100} \qquad [6]$$

$$\frac{k_{p-}}{k_{H}} = \frac{k_{C_6H_5X}}{k_{C_6H_6}/6} \times \frac{\%p\text{-product}}{1 \times 100} \qquad [7]$$

k_{m-}/k_H and k_{p-}/k_H are generally referred to as *partial rate factors* and written as f_{m-} and f_{p-} respectively.

The main bar to testing this extension of Hammett plots into aromatic substitution reactions was initially the lack of enough, sufficiently precise data. As more became available, relatively crude checks suggested that the correlation of partial rate factors with σ_{m-} was not unreasonable, but that there were major deviations with some σ_{p-} values. This is what might have been expected, only more so, by analogy with the behaviour on solvolysis of p-substituted cumyl halides (p. 185); for with the latter a positive charge is developed in the intermediate (and transition state) on a carbon atom adjacent to the benzene nucleus (24) which can be delocalized into the nucleus (24a) itself, while in aromatic substitution the positive charge is developed directly on a nuclear carbon atom (25):

It could be, because of this similarity of intermediate (and transition state), that the solvolysis of cumyl chlorides might thus constitute a better standard reaction than the ionization of $ArCO_2H$ in providing effective values of σ for correlating aromatic electrophilic substitution data. H. C. Brown and his colleagues determined solvolysis rates for a variety of substituted cumyl chlorides, then plotted the log ks for selected m-substituted halides against the most accurately available σ_{m-} values (calculated from thermodynamic ionization constants of m-$XC_6H_4CO_2H$), the resultant straight line gave ρ for the solvolysis (in 90 per cent aqueous acetone) as -4.54. This could then be used, in conjunction with the relevant log ks, to calculate new σ values for a range of p-substituents. These new values are generally written as σ^+ to distinguish them from the σ values obtained from $ArCO_2H$ ionization, the $+$ sign indicating that they apply to reactions in which electron-deficiency (positive charge) is developed at the reaction centre in the transition state. These σ^+ values could then, in turn, be plotted against experimentally determined partial rate factors (k_{m-}/k_H and k_{p-}/k_H) for aromatic electrophilic substitution reactions, and correlations were, on the whole, found to be fairly good. As might have been expected, ρ values for such reactions were found to have large negative values: that for bromination of C_6H_5X in aqueous acetic acid at 25° being $-12\cdot1$.

σ^+ Values—hardly surprisingly—also provide better correlations than do σ values for the side-chain reactions of *m*- and *p*-substituted benzene derivatives in which some degree of $+$ve charge is generated in the transition state, e.g. the solvolysis of halides of the type $ArCR_2Cl$. The ρ value for each series of halides, obtained from the slope of a log $(k/k_0)/\sigma^+$ plot, gives a measure of the relative sensitivity of each reaction to substituent effects:

Chloride	ρ
$ArCPh_2Cl$	$-2\cdot34$
$ArCHPhCl$	$-4\cdot05$
$ArCMe_2Cl$	$-4\cdot54$
$ArCH_2Cl$	$-2\cdot20$

The first and last members of the series are found, rather interestingly, to be much less sensitive to substituent effects than the two intermediate compounds. With $ArCPh_2Cl$ this presumably reflects the fact that in the transition state, which will very closely resemble an ion pair (26), delocalization of the $+$ve charge through the agency of three aromatic nuclei is already so effective that the added effect of X is really neither here nor there:

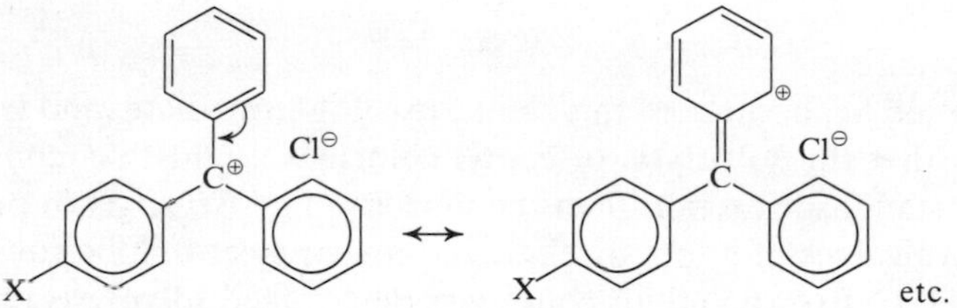

(26)

With $ArCH_2Cl$ the lower substituent effect presumably reflects the fact that its solvolysis is near a mechanistic borderline (greater S_N2 character than the others), so that the development of $+$ve charge in the transition state is relatively less important to the overall reaction.

DUAL σ VALUES

The question thus arises whether any real, or physically valid, purpose is served by adopting dual σ values for a substituent, depending on whether there is, or is not, significant, direct conjugative interaction between the substituent (usually *p*-) and the reaction centre in the transition state. The occurrence of such conjugative interaction would require a σ^+ value (see above) for the direct interaction of an electron-donating substituent with developing positive charge at the reaction centre, and a σ^- value, (*cf.* the effect of p-NO_2 on the ionization of phenols and anilinium cations, p. 183) for the direct interaction of an electron-withdrawing substituent with developing negative charge at the reaction centre.

For a single σ^+ or σ^- value adequately to define a particular substituent's direct conjugative effect implies that the magnitude of that effect will be the same no matter what the reaction, i.e. that the electronic demand on a particular substituent by the reaction centre in the transition state will always be the same. This, on the face of it, sounds inherently unlikely, and doubts grew when it was realized that a σ^- value for p-NO_2 of 1·00 would be required for the ionization of thiophenols, i.e. (27),

(27)

compared with a σ^- value of 1·27 for the ionization of phenols and anilinium cations (p. 183), and a σ value of 0·778 for reactions (e.g. the ionization of $ArCO_2H$) in which conjugative interaction of reaction centre and substituent was believed to be essentially absent.

It is interesting, though not necessarily significant, that the σ^- thiophenol value is more or less midway between the other σ^- value and the σ value for p-NO_2. A definitive verdict on the validity of dual σ values, i.e. σ and σ^-, or σ and σ^+, was provided by Wepster who analysed a great mass of data and found that the required σ values for p-NO_2, far from clustering around 0·778(σ) and 1·27(σ^-), are spread more or less evenly between 0·6 and 1·4. Similarly, required σ values for p-OMe are found not to concentrate round two fixed points but to be spread fairly evenly between 0 and $-0·8$.

The failure of dual σ values adequately to correlate the available data is not perhaps the loss that it at first seems; for what is most interesting about Hammett and kindred plots is not so much the straight line relationships, but deviations from them. In the cases referred to above, the extent of the deviation from the straight line in a particular reaction of, for example, p-NO_2 or p-OMe substituted compounds supplies useful information about the magnitude of the conjugative interaction between the substituent and the developing charge at the reaction centre in the transition state: which in turn supplies significant inferential evidence about the charge distribution in the latter.

Nevertheless, attempts have been made to devise linear free energy relationships that can accommodate this variation of conjugative interaction from one reaction to another. The best example of these is the Yukawa–Tsuno equation

$$\log \frac{k}{k_0} = \rho[\sigma + r(\sigma^+ - \sigma)] \qquad [8]$$

that relates to electron-withdrawing substituents and the development of electron density at the reaction centre in the transition state. Here r is a measure of the direct conjugative requirement of the reaction centre, and is found to have values lying between 2·3 and 0·2 for reactions whose ρ values lie between -12 and $-0·6$. It may be evaluated for a particular reaction by first determining ρ from $\log k$ measurements on m-substituted compounds (here $\sigma_{m^-}{}^+ \approx \sigma_{m^-}$, and r is thus eliminated), then determining r from known σ and σ^+ values, and measured $\log k$ values, for p-substituted compounds. This relation has been useful in correlating more exactly a wider range of data, and the parameter r has the appeal of reflecting a real physical effect: the variation, from one reaction to another, of the conjugative demand on a substituent by the reaction centre. On the other hand, the very introduction of an additional parameter, r, is bound to result in the better correlation of a large mass of related experimental data.

NON-AROMATIC COMPOUNDS—STERIC EFFECTS

When Hammett plots of $\log k/k_0$ against σ are extended to the reactions of aliphatic compounds—and to those of o-substituted benzene derivatives—straight line relationships no longer result, i.e. linear free energy relationships involving the normal polar σ (or σ^+/σ^-) constants no longer apply. This failure is due in part to the different operation of polar effects—than with m-, and particularly, p-substituents—but also involves bulk steric and proximity effects—the latter particularly with aliphatic compounds which are more flexible than the rigid aromatic derivatives—operating on the reaction centre, or on the attacking reagent, e.g. hydrogen bonding, reagent complexing, neighbouring group participation (*cf.* p. 133) etc.

The major attempt to establish linear free energy correlations over a wider field has been made by Taft. Acting on a suggestion originally made by Ingold, Taft compared the relative susceptibility to substituent effects ($\log k/k_0$ against σ) of the hydrolysis of (aromatic) esters under acid- and base-catalysed conditions, respectively. He found that base-catalysed hydrolysis (and trans-esterification) was relatively highly susceptible to polar substituent effects ($\rho \approx 2\cdot5$), while acid-catalysed hydrolysis (and esterification) was very little susceptible ($\rho \approx 0$). Bearing in mind the close, formal similarity of the transition states for acid-catalysed (28) and base-catalysed (29) hydrolysis

$$\left[\begin{array}{c} \text{OH} \\ \| \\ \text{R}\!-\!\text{C}\cdots\text{OH}_2 \\ | \\ \text{OR}' \end{array} \right]^{\oplus} \qquad \left[\begin{array}{c} \text{O} \\ \| \\ \text{R}\!-\!\text{C}\cdots\text{OH} \\ | \\ \text{OR}' \end{array} \right]^{\ominus}$$

$$(28) \qquad\qquad\qquad (29)$$

it seems not unreasonable to assume that any changes in steric and conjugative effects between starting material and transition state will be closely similar for the two processes. The substituent effect on the base-catalysed reaction will thus be made up of polar (P), conjugative (C) and steric (S) effects

$$\log\left(\frac{k}{k_0}\right)_{\text{BASE}} = \text{P} + \text{C} + \text{S} \qquad [9]$$

while the effect on the acid-catalysed reaction will be made up of essentially the same conjugative and steric effects, the polar effect here being negligible (see above):

$$\log\left(\frac{k}{k_0}\right)_{\text{ACID}} = \text{C} + \text{S} \qquad [10]$$

Subtraction of [10] from [9] thus evaluates the polar effect (P), and if this is expressed in terms of substituent (σ^*), and reaction (ρ^*), constants similar to, but different from, the familiar Hammett terms

$$\log\left(\frac{k}{k_0}\right)_{\text{BASE}} - \log\left(\frac{k}{k_0}\right)_{\text{ACID}} = \text{P} = \rho^*\sigma^* \qquad [11]$$

we then have a relation [11] which may have a wider validity. Thus alkyl ester hydrolysis, $\log k$ for RCO_2R' ($\log k_0$ for CH_3CO_2R') or for $o\text{-}RC_6H_4CO_2R'$ ($\log k_0$ for $o\text{-}CH_3C_6H_4CO_2R'$), was chosen as the standard reaction but ρ^* was specified not as $1\cdot00$ (*cf.* $\rho = 1\cdot00$ for the Hammett standard reaction, the ionization of *m*- or *p*-substituted $ArCO_2H$, p. 171) but as $2\cdot48$ in order to yield σ^* values not too numerically different from the familiar σ values (the reference substituent is, of course, now CH_3 and not H). The simple Taft relation can then be used in the more general form

$$\log \frac{k}{k_0} = \rho^*\sigma^* \qquad [12]$$

to evaluate ρ^* for further reactions, and to extend the range of σ^* values to other substituents.

The degree of success in the correlation of reaction data achieved by use of the above relation is essentially a measure of how far, in a reaction, the polar effect only is operative. Deviations from a straight line relationship are a measure of the degree of involvement of conjugative (of relatively little significance with saturated aliphatic compounds) and general steric effects. Attempts to evaluate the latter were made by Taft using the relation for acid-catalysed ester hydrolysis

$$\log \left(\frac{k}{k_0} \right)_{ACID} = C + S \qquad [10]$$

for cases in which the conjugative effect (C) was believed to be very small or non-existent; it was thus possible to evaluate a steric parameter, E_s, for a number of substituents from the relation:

$$\log \left(\frac{k}{k_0} \right)_{ACID} = E_s \qquad [13]$$

It will be seen from the way that E_s is defined (k will be smaller than k_0 if the reaction rate of the R-substituted compound is slowed by steric inhibition) that bulkier groups than CH_3 (the standard) will have negative E_s values: the larger the group, the larger the negative value. Thus for o-substituted aromatic derivatives, nitro—hardly surprisingly—qualifies as a large group, methyl and bromine as much smaller and about the same size as each other; while chlorine is slightly smaller, and iodine rather larger, than they are: all as expected. Simple alkoxy groups qualify as being very small, however, which is contrary to experience, and the reason for this is not clear. Purely steric correlation of reaction rates is, as expected, most successful for alkyl groups for with them polar effects are known to be minimal.

It is now possible to set up a relation incorporating both polar and steric effects

$$\log \frac{k}{k_0} = \rho^*\sigma^* + \delta E_s \qquad [14]$$

where δ is a parameter indicating the reaction's sensitivity to steric effects; it is made equal to $1\cdot00$ for ester hydrolysis at 25°. δ can then be evaluated, for a particular reaction, in the normal way: for example, a value of δ of $0\cdot81 \pm 0\cdot03$ was obtained for the acid-catalysed hydrolysis of *o*-substituted benzamides (o-$RC_6H_4CONH_2$). This extended Taft relation is obviously an advance on a simple Hammett relation in that it correlates a less restricted field of reactions, and incorporates specific polar and steric parameters. The very increase in the number of parameters employed, however, must needs make for better correlation of reaction data and, in any case, as with the simple Hammett relation, it is usually deviations (rather than neat straight lines!) that are most interesting: indicating as they do the intervention of unexpected effects that have not yet been allowed for. Thus major 'steric' deviations from Taft plots are often indicative of neighbouring group participation, either by interaction with the reaction centre, or by complexing of substituents with the attacking reagent.

SOLVENT EFFECTS

One of the most profound factors influencing equilibrium constants or reaction rates is the medium—the solvent—in which the process is taking place: in our considerations to date no direct attempt has been made to take such effects into account. It is of course true that the ρ or ρ^* (reaction constant) values of Hammett or Taft relations, respectively, will change if the solvent in which a particular reaction is taking place is changed. Thus while the ρ value for the ionization of $ArCO_2H$ at 25° in water is, by definition, 1·00, the value in ethanol is found to be 1·957, at the same temperature; similarly, the ρ value for the base-catalysed hydrolysis of $ArCO_2Et$ at 25° in 70 per cent aqueous dioxan is found to be 1·828, while in 85 per cent aqueous ethanol the value is 2·537. This is not, however, the same as seeking to correlate reactivity directly with medium and, by implication, being able to make (useful) predictions about the best solvent in which to carry out a particular reaction.

Early attempts were made to relate the varying reactivity of a substrate in a range of different solvents to some readily measurable, macroscopic, property of the solvent. For there to be any real hope of success a property has to be chosen that has some apparent, general relevance to the known mechanistic character of the reaction being considered: in particular to significant changes between starting material and transition state. A good example is a reaction in which there is a separation of charge in the rate-limiting stage, e.g. the S_N1 reactions of halides, in which the transition state is thought to resemble the ion-pair intermediate (30),

$$R-X \xrightarrow{k_1} R^{\oplus}X^{\ominus}$$

(30)

where it might be worth seeking to relate reaction rate to the dielectric constant of the solvent, ε. What is observed in this particular case is that, in general terms, the more polar the solvent—the larger the value of ε—the faster is the S_N1 solvolysis of a particular halide, provided the solvent remains of the same general type. Thus there is a rough qualitative relation of k to ε for a series of alcohols, but this no longer holds if one of the alcohols in the sequence is replaced by, for example, a nitrile having the same ε value. Even in a solvent series having a common functional group there is no straight line relationship between ε and log k or anything remotely approaching one. This is hardly surprising for in the above reaction the ion-solvating ability of the solvent will clearly be of importance, in addition to its dielectric constant, and we have not sought to take any account of it. There is no necessary relation between these two properties, as seen with hydroxylic solvents such as the alcohols which have a far greater ion-solvating ability than other non-hydroxylic solvents of higher dielectric constant.

A similar fate has overtaken other attempts to relate the varying rate of reaction of a particular substrate in a series of solvents to a single macroscopic property of the solvent series. An attempt at the now familiar empirical establishment of parameters in a linear free energy relationship—followed only subsequently by their justification in familiar physical terms—was made by Grunwald and Winstein for the S_N1 solvolysis of Me_3CCl, chosen arbitrarily but conveniently as the standard substrate,

$$\log k^{\circ}_{A} - \log k^{\circ}_{B} = Y_{A} - Y_{B} \qquad [15]$$

where k_{A}° and k_{B}° are the specific rate constants for the solvolysis of Mc_3CCl in solvents A and B, respectively, and Y_{A} and Y_{B} are solvent constants. Selecting the solvent 80 per cent aqueous ethanol as the standard (B) and making Y_{B} equal to zero, allows the experimental evaluation of a series of Y values for other solvents:

Solvent	Y
Water	+3·493
80% Aq. ethanol	0·000 (by definition)
80% Aq. acetone	−0·673
80% Aq. dioxan	−0·833
Methanol	−1·090
Ethanol	−2·030

A general—and recognizable—relation for the S_N1 solvolysis of any substrate can then be established

$$\log \frac{k_A}{k_0} = m\mathbf{Y}_A \qquad [16]$$

where k_A and k_0 are the first order solvolysis rate constants for the particular substrate in a solvent A and in the standard solvent (80 per cent aqueous ethanol), respectively, m is the compound parameter ($m = 1{\cdot}00$ for Me_3CCl) measuring the sensitivity of the rate of solvolysis to $\mathbf{Y}_A$, which is itself a measure of the 'ionizing power' of the solvent A.

We are now back on relatively familiar territory, and the question arises how successful is the Grunwald–Winstein relation at correlating data and, perhaps more importantly, how useful is it, in terms of the relative magnitude of m and $\mathbf{Y}$ values and in deviations from linearity, at providing useful mechanistic information. In general the correlations are quite good over fairly restricted areas: this is as much as can reasonably be expected for a two-parameter relation covering a wide variety of substrates, solvents and, almost certainly, mechanistic types; it is in this last respect that Grunwald–Winstein plots can prove particularly interesting.

Uni- and bi-molecular solvolysis reactions are not normally readily distinguishable by simple kinetic means for though the solvent is directly involved in the rate-limiting stage of the latter, i.e. it is a constituent of the transition state (31),

$$\overset{\delta+}{S}\cdots R \cdots X^{\delta-}$$
$$(31)$$

it is present in such excess that its concentration does not vary significantly; the reaction is thus pseudo-unimolecular, and indistinguishable kinetically from the truly uni-molecular mode. Stereochemical evidence can be used but this may be laborious and is not always decisive (p. 128). In the parameter, m, however we have an index of a substrate's sensitivity, in solvolysis, to the 'ionizing power' of the solvent. A major difference between S_N1 and S_N2 transition states is that the former closely resembles an ion-pair (30), in which unit charge is essentially wholly developed, while in the latter (31), only partial charges have formed and they are much less concentrated, being spread over a more extended species. We should thus expect the former, S_N1, to be much the more sensitive of the two to solvent 'ionizing power'—as we know in fact to be the case—and thus to be characterized by larger values of m. In m we may thus have a useful diagnostic tool:

Halide	Temp.	m value
Me_3CCl	25°	1·00
Me_3CBr	25°	0·94
$EtMe_2CBr$	25°	0·90
$PhCH(Me)Br$	50°	1·20
$CH_2{=}CHCH(Me)Cl$ ·	25°	0·89
$EtBr$	75°	0·34
$Me(CH_2)_3Br$	75°	0·33

An m value larger than 1·00 implies that the substrate is more sensitive to solvent 'ionizing power' than is Me_3CCl, i.e. that the transition state to which it gives rise is more 'ion pair like' in character. The smaller the m value the less ion pair like in character will be the transition state, and the purely bimolecular (S_N2) solvolyses of bromoethane and 1-bromobutane are seen to have m values of $\approx 0\cdot3$. In general, an m value of $\approx 0\cdot5$ may be taken as an approximate indicator of the mechanistic borderline: for values above this there is significant, and increasing, ion pair character in the transition state, for values below it there is little or none.

THERMODYNAMIC REQUIREMENTS FOR LINEAR FREE ENERGY RELATIONSHIPS

Log K and log k values are defined by $\Delta G°$ and $\Delta G^{\ddagger}$, respectively, and both these terms involve enthalpy (ΔH) and entropy (ΔS) components:

$$\Delta G° = \Delta H° - T\Delta S° \qquad [17]$$

$$\Delta G^{\ddagger} = \Delta H^{\ddagger} - T\Delta S^{\ddagger} \qquad [18]$$

In plots of log K against log K' (Fig. 5.2, p. 170) or of log K against log k (Fig. 5.1, p. 170) (or, of course, of log k against log k'), we are thus plotting the free energy change values for a series of compounds undergoing one reaction against the free energy change values for a different series of compounds undergoing another reaction. Given that each ΔG term is made up of two different, and independently variable, functions (ΔH and ΔS) we would expect a straight line relationship between the two series to result only if, for each series, one or other of the following conditions holds:

1) ΔS is constant for the series.
2) ΔH is constant for the series.
3) ΔH is linearly related to ΔS for the series.

These are extremely stringent conditions and there is considerable doubt whether they are in fact met by many of the reaction series that show good correlations in Hammett plots: there is indeed doubt whether they are met even by the ionization of $ArCO_2H$ in water at 25° series, the Hammett standard reaction! The fact that *linear* free energy relationships are indeed observed, nevertheless, no doubt arises in some cases from mutual cancelling out of divergent ΔH and ΔS trends, but this cannot be the explanation for all the observed linear correlations. Further and more detailed theoretical analysis of the influence of polar substituents on the energy (potential and kinetic) of reacting systems is beginning to provide some insight, but not all is yet clear and linear free energy relationships remain a testament to the theoretical utility of concepts that are empirical in their genesis.

FURTHER READING

CHAPMAN, N. B. and SHORTER, J. (Eds.). *Advances in Linear Free Energy Relationships* (Plenum Press, 1972).

CHARTON, M. 'The Quantitative Treatment of the Ortho Effect', *Progr. Phys. Org. Chem.* (Interscience), 1971, **8**, 235.

EHRENSON, S. 'Theoretical Interpretation of the Hammett and Derivative Structure-Reactivity Relationships', *Progr. Phys. Org. Chem.* (Interscience), 1964, **2**, 195.

HAMMETT, L. P. 'Some Relations between Reaction Rates and Equilibrium Constants', *Chem. Rev.,* 1935, **17**, 125.

HAMMETT, L. P. *Physical Organic Chemistry* (McGraw-Hill, 2nd Ed. 1970), pp. 347–408.

JAFFÉ, H. H. 'A Re-examination of the Hammett Equation', *Chem. Rev.,* 1953, **53**, 191.

JAFFÉ, H. H. and LLOYD JONES, H. 'Applications of the Hammett Equation to Heterocyclic Compounds', *Adv. in Heterocyclic Chem.* (Academic Press), 1964, **3**, 209.

LEFFLER, J. E. and GRUNWALD, E. *Rates and Equilibria of Organic Reactions* (Wiley, 1963), pp. 128–402.

RITCHIE, C. D. and SAGER, W. F. 'An Examination of Structure-Reactivity Relationships', *Progr. Phys. Org. Chem.* (Interscience), 1964, **2**, 323.

SHORTER, J. 'Linear Free Energy Relationships', *Chem. in Britain,* 1969, **5**, 269.

SHORTER, J. 'The Separation of Polar, Steric and Resonance Effects in Organic Reactions by the Use of Linear Free Energy Relationships', *Quart. Rev.,* 1970, **24**, 433.

SHRECK, J. O. 'Non-linear Hammett Relationships', *J. Chem. Ed.,* 1971, **48**, 103.

STOCK, L. M. and BROWN, H. C. 'A Quantitative Treatment of Directive Effects in Aromatic Substitution', *Adv. Phys. Org. Chem.* (Academic Press) 1963, **1**, 35.

TAFT, R. W. 'Separation of Polar, Steric, and Resonance Effects in Reactivity', *Steric Effects in Organic Chemistry,* Ed. Newman, M. S. (Wiley, 1956), pp. 556–675.

WELLS, P. R. 'Linear Free Energy Relationships', *Chem. Rev.,* 1963, **63**, 171.

WELLS, P. R. *Linear Free Energy Relationships* (Academic Press, 1968).

6
Some further examples

The idea behind this chapter is to provide a few additional, and contrasted, examples in which the techniques and arguments that have been employed previously are either carried a little further, or several of them are used together, to provide somewhat more sophisticated information about the pathway by which a particular reaction is believed to proceed.

KINETIC EVIDENCE FOR REACTIVE INTERMEDIATES*

This interesting and ingenious example postulates the involvement of a reactive intermediate—not detectable by physical or other methods—in a Diels Alder reaction, and then confirms the actual occurrence of such an entity by comparing the detailed kinetic model that its involvement would entail with what is actually observed experimentally.

*R. Huisgen, F. Mietzsch, G. Boche and H. Seidl, *Organic Reaction Mechanisms*, Special Publication No. 19 (Chem. Soc., London), 1965, pp. 3—20.

(a) The reaction

The reaction in question is the addition of dienophiles, such as tetra-cyanoethylene (1), to cyclooctatetraene (2) to yield adducts which have been shown, on the basis of their n.m.r. spectra and other evidence, to have structures such as (3):

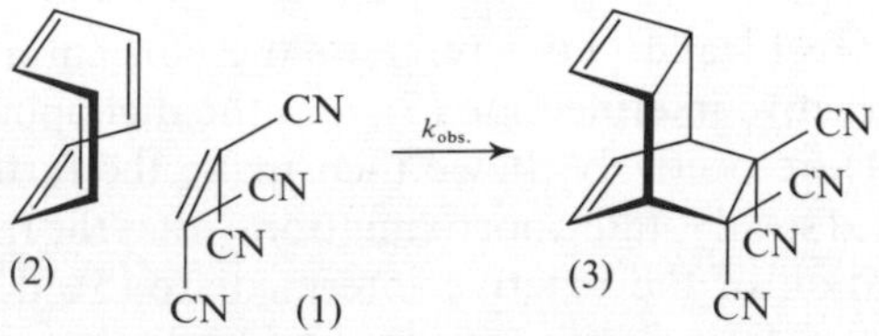

The above representation of the addition to yield (3) is unsatisfactory, however, in that though the adduct can, purely formally, be obtained by addition of the dienophile across the two separated double bonds in (2), these bonds are not in fact conjugated with each other as the Diels–Alder reaction would require them to be. Further, the normal requirement is not only that the double bonds in the diene should be conjugated, but also that they should be able to assume a *cis*-planar conformation as, for example, in (4);

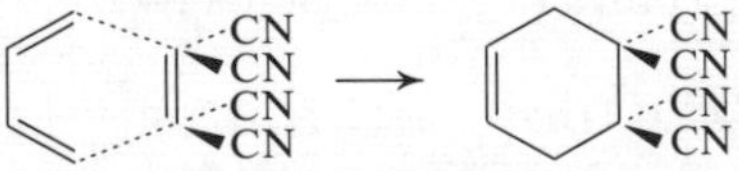

a state that is not attainable by any contiguous pair of double bonds in the non-planar, tub-shaped structure (2). It was therefore suggested, on the basis of the known structure of the adduct (3), that the addition of a dienophile to (2) is preceded by the latter's tautomerization to an intermediate, such as bicyclo[4.2.0]octa-2,4,7-triene (5), which possesses the required structural features: the cyclohexadiene ring in (5) is now essentially flat, so that its double bonds are both conjugated and in the *cis*-planar conformation:

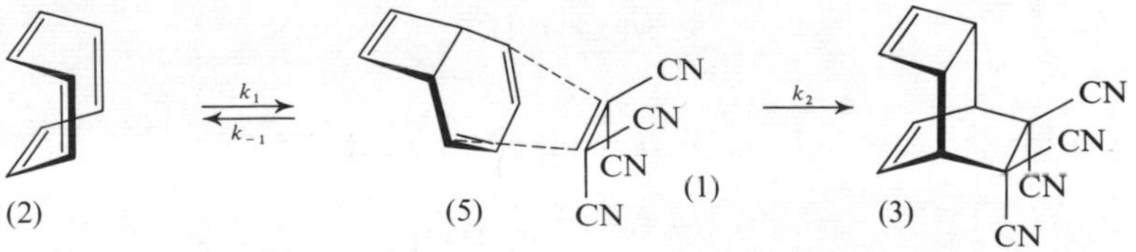

(b) The theoretical model

Assuming that the reaction follows such a pathway, we can write the equation [1] for the rate of formation of the adduct (3),

$$\frac{d[ADD]}{dt} = k_2[INT][DP] \tag{1}$$

where [ADD], [INT] and [DP] refer to the concentrations of the adduct (3), the reactive intermediate (5), and the dienophile, i.e. tetracyanoethylene (1), respectively. If we then make the further assumption—the so-called steady state approximation—that the rate of change of the concentration of the reactive intermediate (5), $d[INT]/dt$, is either zero or essentially insignificant compared with the rates of change of concentration of the other reactants or products, then we can write,

$$\frac{d[INT]}{dt} = k_1[DN] - k_{-1}[INT] - k_2[INT][DP] = 0 \tag{2}$$

where [DN] refers to the concentration of the diene, cyclooctatetraene (2). From [2] it follows that,

$$[INT] = \frac{k_1[DN]}{k_{-1} + k_2[DP]} \tag{3}$$

and incorporating [3] into [1] thus yields [4]:

$$\frac{d[ADD]}{dt} = \frac{k_1 k_2[DN][DP]}{k_{-1} + k_2[DP]} \tag{4}$$

The rate of reaction was actually measured dilatometrically (*cf.* p. 9; there is in fact a contraction in volume), using sufficient excess of dienophile (1) for the reaction to be pseudo first order:

$$\frac{d[ADD]}{dt} = k_{obs.}[DN] \tag{5}$$

Equating the theoretical model [4] with the experimentally established rate law [5] is seen to require that $k_{obs.}$ should be a function of [DP], the concentration of the dienophile (1):

$$k_{obs.} = \frac{k_1 k_2[DP]}{k_{-1} + k_2[DP]} \tag{6}$$

Equation [6] can, in turn, be regrouped to yield [7],

$$k_{obs.} = k_1 - \frac{k_{-1}}{k_2}\frac{k_{obs.}}{[DP]} \tag{7}$$

which is the equation of a straight line for a plot of $k_{obs.}$ against $k_{obs.}/$[DP].

(c) The fit of experiment with theory

Measurement of $k_{obs.}$ for a series of values of [DP], at constant temperature, and plotting of $k_{obs.}$ against $k_{obs.}/$[DP] was indeed found to yield a straight line, whose slope must correspond to k_{-1}/k_2 and whose intercept will give the value of k_1. Similar straight lines could also be obtained for the same dienophile, i.e. tetracycanoethylene (1), at different temperatures, and also for a series of other dienophiles, e.g. maleic anhydride etc. The equilibrium concentration of the intermediate in dioxan at 100° has been shown, indirectly from kinetic data, to be ≈ 0.01 per cent.

The involvement of an intermediate is thus demonstrated unequivocally as a result of this observed fit of experimental kinetic data with the detailed requirements of the theoretically derived model, but this fact does not, of course, provide us of itself with any further information about the actual nature of such an intermediate. There is, however, a closely analogous reaction in which cycloocta-1,3,5-triene (6) undergoes attack by dienophiles, e.g. tetracyanoethylene (1), to yield adducts such as (7), essentially analogous to (3). Here, too, an intermediate is involved, but in this case it is possible to isolate it (it is stable below 60°), and to show it has the structure (8), i.e. that it is bicyclo[4.2.0]octa-2,4-diene:

It has further been shown in this case that it is (8) that is converted into the adduct (7), (6) not reacting with dienophiles directly under the conditions of the reaction but only *via* (8); the fact that (8) can actually be obtained pure allows the several rate constants, analogous to those above, to be separately evaluated. The identification of (8), coupled with the fact that the kinetics of the (6) ⟶ (7) conversion closely parallel those for the (2) ⟶ (3) conversion, provide substantial evidence that the intermediate in the latter conversion does indeed have the structure (5) suggested initially. Clinching evidence for the validity of (5), and of its relationship to cyclooctatctraene (2), has subsequently been provided by its preparation in ≈ 95 per cent yield (at $-78°$) by

the debromination of (9),

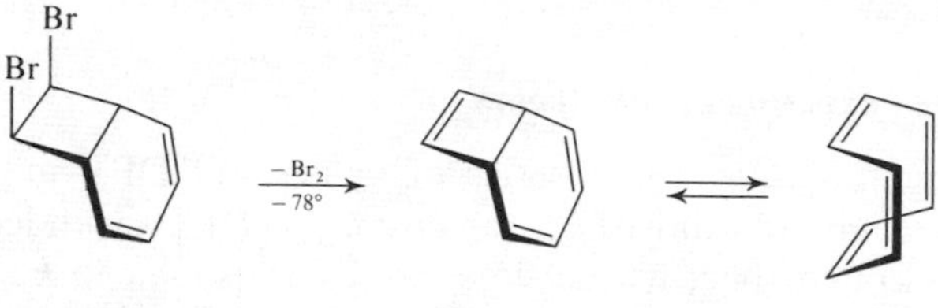

(9)   (5)   (2)

and by the observation that it is converted into (2) even at temperatures below 0°. The half-life of (5), with respect to (2), has been shown to be ≈ 14 min. at 0°.

THE REDUCTION OF THIAZOLIUM SALTS WITH SODIUM BOROHYDRIDE*

An example of the use of deuterium (from D_2O and $NaBD_4$) to chart the pathway followed by a reaction is seen in the reduction, with sodium borohydride in hydroxylic solvents, of N-alkylthiazolium salts, e.g. (1), to the corresponding tetrahydro-derivatives, N-alkylthiazolidines (2):

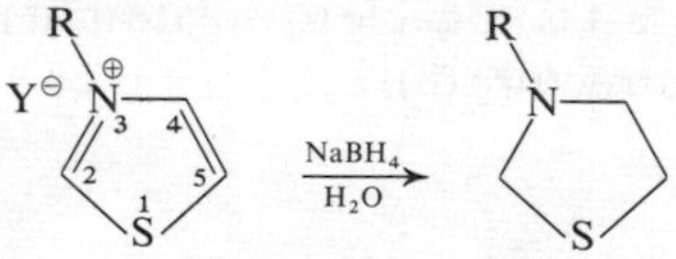

(1)   (2)

(a) The reaction pathway

It seems likely that initial attack is by the highly nucleophilic $BH_4^{\ominus}$ on the 2-position of (1) to yield the dihydro-derivative, an N-alkyl-4-thiazoline (3),

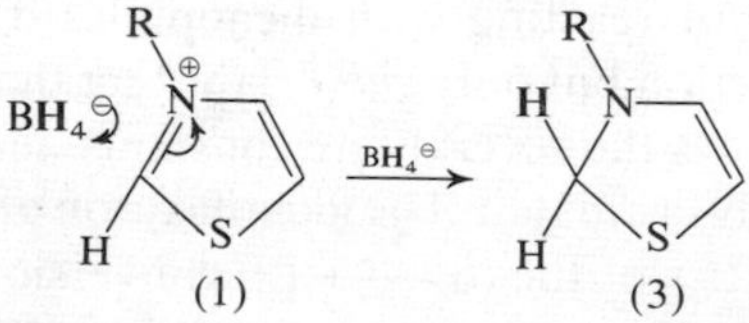

(1)   (3)   [1]

as this position is known to be the site attacked by other nucleophiles, e.g. $^{\ominus}OH$. Further confirmation is provided by the fact that reduction with $NaBH_4$ in non-hydroxylic solvents is found to proceed as far as

*G. M. Clarke and P. Sykes, *J. Chem. Soc.* (C), 1967, 1269, 1411.

the dihydro-derivative (3), but no farther. This apparent need of a hydroxylic solvent to effect reduction beyond this stage suggests that the next step probably involves such solvent, presumably as a proton donor:

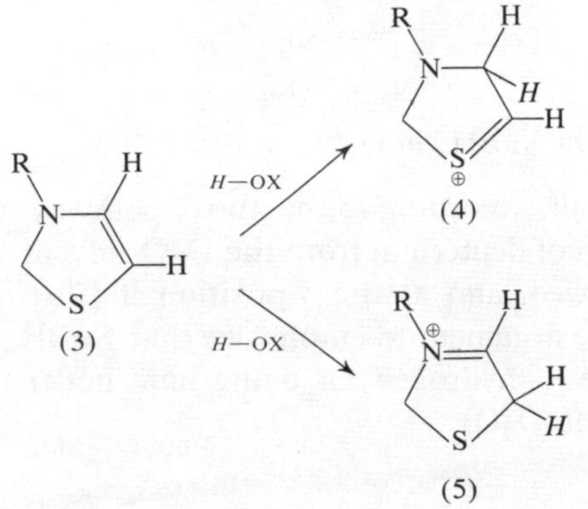

[2a]

(4)

(3) $H-OX$

[2b]

(5)

Such uptake of proton could, in theory, take place at either end of the double bond, i.e. at the 4-position to yield (4) *or* at the 5-position to yield (5), though the latter is much the more likely. This would then be followed by further attack on the resultant cation to yield the tetrahydro-derivative (2):

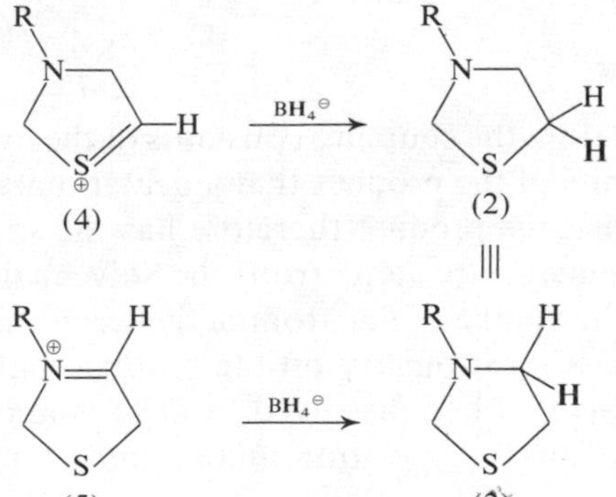

(4) $\xrightarrow{BH_4^{\ominus}}$ (2)   [3a]

‖

(5) $\xrightarrow{BH_4^{\ominus}}$ (2)   [3b]

If such a pathway is indeed followed, then it should be possible to confirm it in detail, including distinguishing between [2a] and [2b] (and consequently between [3a] and [3b]), by carrying out three experiments (the first two are all that are really necessary, the third is merely an added safeguard):

(i)   Reduction with $NaBH_4$ in D_2O
(ii) Reduction with $NaBD_4$ in H_2O
(iii) Reduction with $NaBD_4$ in D_2O

The experiments were actually carried out on 3-benzyl-4-methyl-thiazolium bromide (1a):

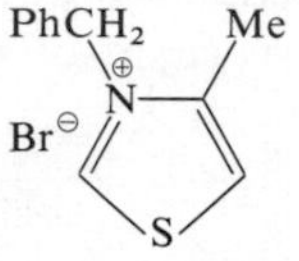

PhCH$_2$ Me

Br$^\ominus$

(1a)

(i) Reduction with NaBH$_4$ in D$_2$O:

This should result, according to the above pathway, in the incorporation of *one* atom of deuterium from the D$_2$O solvent: at the 4-position if [2a] is followed, and at the 5-position if [2b] is followed (it is important to the argument to emphasize that NaBH$_4$ has been shown *not* to exchange its hydrogen for deuterium, under these conditions, when dissolved in D$_2$O):

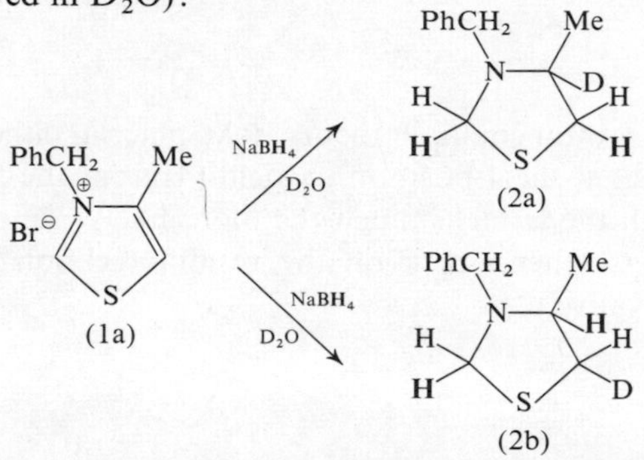

PhCH$_2$ Me

PhCH$_2$ Me   NaBH$_4$ / D$_2$O   (2a)

Br$^\ominus$

(1a)   NaBH$_4$ / D$_2$O

PhCH$_2$ Me

(2b)

It may be shown from the coupling constants of the hydrogen atoms in the n.m.r. spectrum of the product that a deuterium atom has entered the 5-position, that the product therefore has the structure (2b), and the uptake of deuteron (proton) from the solvent thus proceeds *via* pathway [2b]. The uptake of the atom of deuterium can, of course, be monitored by mass spectrometry on the product, but there is a complication in that (1a), when dissolved in D$_2$O is known to exchange the hydrogen atom at its 2-position quite rapidly for deuterium (1b):

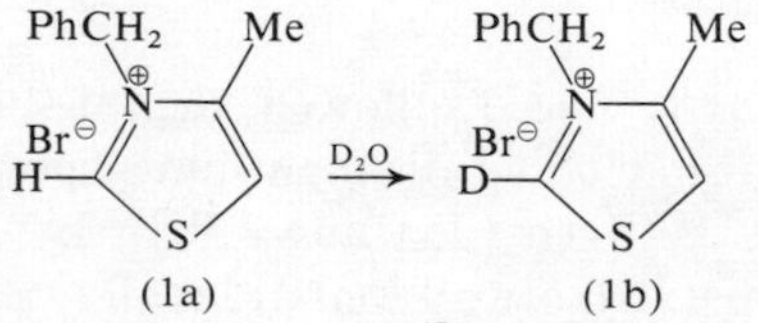

PhCH$_2$ Me   PhCH$_2$ Me

Br$^\ominus$ Br$^\ominus$
H—   D$_2$O   D—

(1a)   (1b)

The n.m.r. and mass spectra of product (2b) thus show, on average, the incorporation of between slightly more than one up to as much as two atoms of deuterium, depending on how soon NaBH$_4$ is added after (1a) has been dissolved in the D_2O (each individual molecule will, of course, contain either one *or* two atoms of deuterium); no exchange can occur once reduction has taken place. This merely results in an added complication, however, and constitutes no bar to the validity of the argument.

(ii) Reduction with NaBD₄ in H₂O:

This should result in the incorporation of *two* atoms of deuterium, one at the 2-position through [1] and one at the 4-position through [3b]:

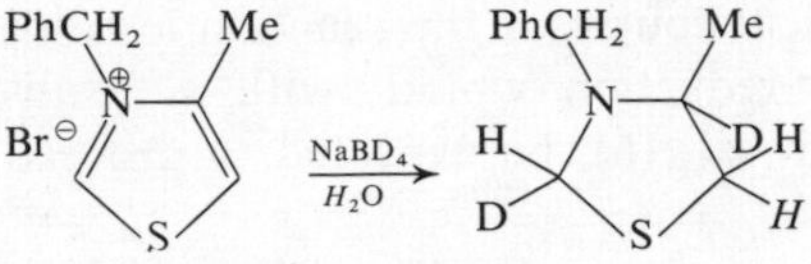

(1a)   (2c)

It may be shown by mass spectrometry that two atoms of deuterium have been incorporated into (2c) from the reducing agent, $NaBD_4$; that one of them has entered the 4-position is confirmed by comparison of the n.m.r. spectrum of (2c) with that of the all-hydrogen thiazolidine (2d),

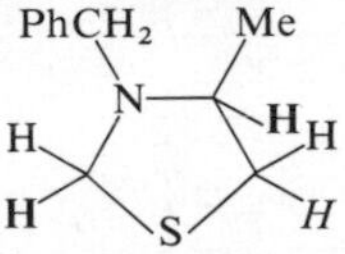

(2d)

when it is found that the three hydrogen doublet, arising from the methyl group in the 4-position of (2d),

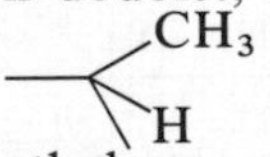

is replaced by a three hydrogen singlet, arising from the methyl group in the 4-position of (2c),

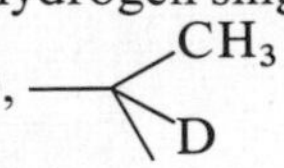

(iii) Reduction with NaBD₄ in D₂O:

This should, of course, result in the incorporation of *three* atoms of deuterium, those at the 2- and 4-positions coming from the $NaBD_4$ *via* [1] and [3b], respectively, and that at the 5-position coming from the solvent, D_2O, *via* [2b]:

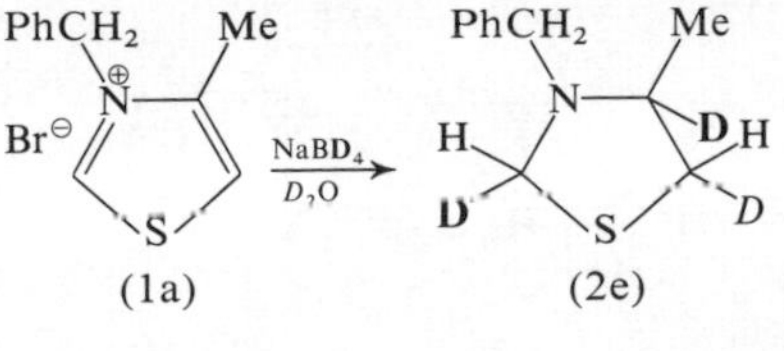

(1a)   (2e)

This incorporation, too, may be confirmed by study of the n.m.r. and mass spectra of (2e), though as in (i) exchange of the hydrogen atom in the 2-position of (1a) with D_2O before reduction results in the observed uptake of deuterium being anywhere between slightly more than three up to as much as four atoms depending on how soon $NaBD_4$ is added after (1a) has been dissolved in the D_2O.

The projected pathway is thus confirmed in detail.

(b) The stereochemistry of reduction

The stereochemical course of the reaction also presents some points of interest; the reduction would, with a 2-substituted-3-benzyl-4-methylthiazolium salt (6), be expected to proceed by the following, stepwise sequence:

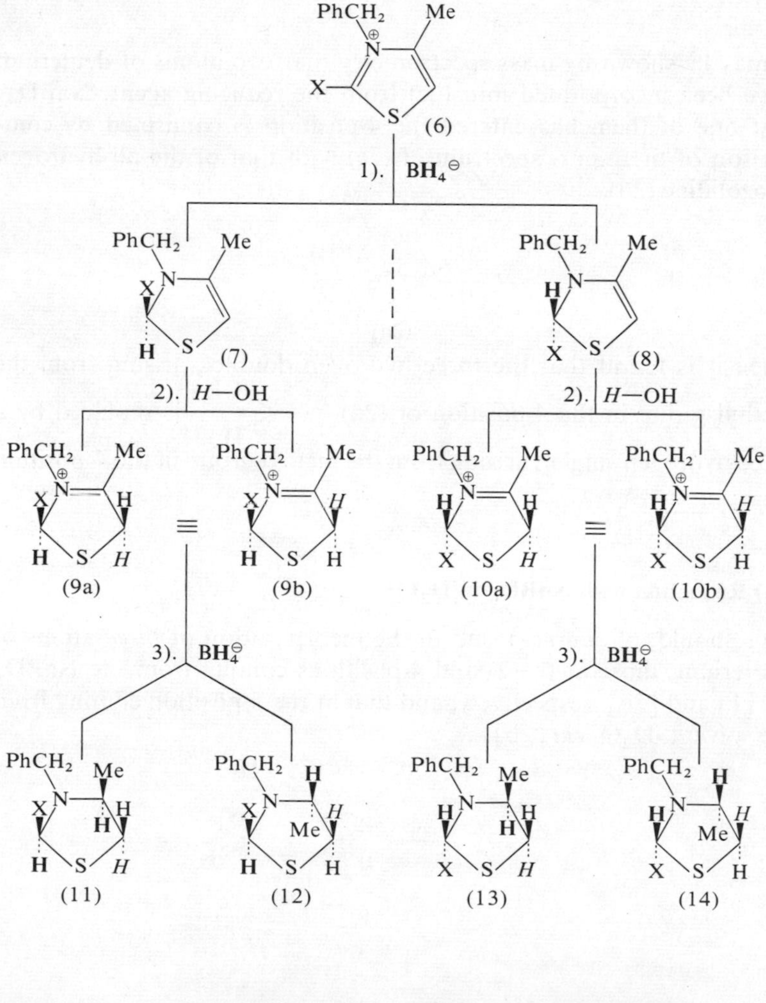

In 1). attack by $BH_4^{\ominus}$ on the 2-position of the essentially planar thiazolium nucleus can take place with equal facility (i.e. at the same rate) either from *below* this plane—to yield (7)—or from *above* this plane—to yield (8): the result is thus a 50:50 mixture of this pair of mirror images, i.e. a racemate ($\pm$). This is then followed, in 2)., by donation of a proton from the solvent, H_2O. This, too, can take place either from below the plane of the ring—to yield (9a)—or from above this plane—to yield (9b); the two faces of the ring are, however, no longer equivalent (because of the X *above* and H *below* at the 2-position in 7), and (9a) and (9b) will therefore be formed at different rates, i.e. there will *not* be equal quantities of them in the reaction product. In this particular case, because the 5-position already carries a hydrogen atom, addition of the second hydrogen atom (actually proton) from above and below the plane of the ring will yield products that are indistinguishable, i.e. (9a) and (9b) are, of course, identical. Thus (7) yields (9a≡b), while (8) yields the mirror image (10a≡b).

In the third, and last, step 3). a second hydrogen atom is being donated by $BH_4^{\ominus}$, this time to the 4-position which already carries a methyl substituent. Attack on (9a≡b) from below and above the plane of the ring will this time, therefore, yield different products—(11) and (12), respectively—and at different rates because of the non-equivalence of the two faces of the ring: the diastereomers, (11) and (12), will thus be obtained in different amounts. Similarly, (10a≡b) will yield different amounts of the diastereomers, (13) and (14), but because of the mirror image relationship of (10a≡b) to (9a≡b), the relative proportion of (14) to (13) will be the same as the relative proportion of (11) to (12). As may be seen on inspection (11) and (14) are mirror images, as are (12) and (13): the end result of the reduction should thus be *two racemates* ($\pm$), obtained in different relative amounts.

The relative proportions of the two racemates obtained is a measure of the degree of stereoselectivity of step 3). above—the final attack by $BH_4^{\ominus}$ on the 4-position—and might be expected to differ depending on the substituent X. For the salt (15), the following results were obtained (the relative proportions of the racemates were normally determined by analysis of n.m.r. spectra):

(15)

	X	R	Rac.1.	Rac.2.
(15a)	D	$PhCH_2$	1	1
(15b)	Me	$PhCH_2$	4	5
(15c)	Ph	Me	1	20
(15d)	$PhCH_2$	$PhCH_2$	0	1

The stereoselectivity of attack in step 3). is thus found to increase, reasonably enough, as the size of the substituent X increases. With (15c) it was possible to obtain one racemate, the major component, pure by recrystallization, while with (15d) only one of the two expected racemates could actually be detected. It might be thought that the preferred racemate would be that in which $BH_4^{\ominus}$ approaches $(9a \equiv b)$ and $(10a \equiv b)$ from the face opposite to the X substituent, i.e. $(11 + 14)$, and this would probably be so if simple kinetic control (*cf.* p. 33) were operative. It might be, however, that the preferred racemate is the more thermodynamically stable one in which X and Me are *trans* to each other, i.e. $(12 + 13)$; it is therefore necessary to identify the configuration of the more abundant racemate by X-ray crystallography or other means: this has not yet been done.

THE FAVORSKII REARRANGEMENT*

Passing reference has already been made (p. 103) to the Favorskii rearrangement in which attack of base, e.g. $^{\ominus}OH$, $^{\ominus}OR$ on an α-haloketone (1) converts it, somewhat unexpectedly, into a carboxylic acid or a derivative thereof (2):

$$RCH_2COCH_2Cl \xrightarrow{\ \ominus OR'\ } RCH_2CH_2CO_2R'$$
$$(1) \hspace{4cm} (2)$$

(a) The nature of the intermediate

Some clue to what may be taking place is afforded by the observation that the isomeric chloroketones (3) and (4) both yield the salt of the same acid (5) on treatment with base,

$$PhCH_2-\overset{\overset{\displaystyle O}{\|}}{C}-\underset{\underset{\displaystyle Cl}{|}}{CH_2} \longrightarrow PhCH_2CH_2CO_2^{\ominus} \longleftarrow PhCH\overset{\overset{\displaystyle O}{\|}}{-C}-CH_3$$
$$\hspace{0.5cm}(3) \hspace{3.5cm} (5) \hspace{3cm} (4)$$

which suggests the involvement of a common intermediate derivable from both (3) and (4). The suggestion was made that this might be the

---

Cf. F. G. Bordwell, *Accounts of Chem. Res.*, 1970, **3**, 286; D. J. Cram, *Fundamentals of Carbanion Chemistry* (Academic Press, 1965), pp. 243–249.

cyclopropanone derivative (6),

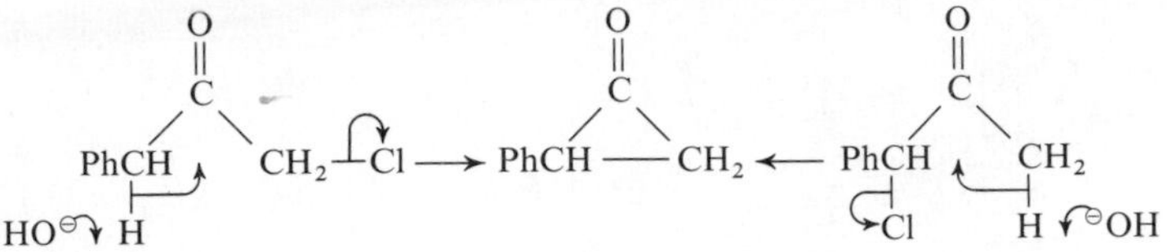

(3)   (6)   (4)

which would then be expected to undergo ready ring-opening by base to yield the more stable of the two possible carbanion intermediates, i.e. (7a) rather than (7b)*:

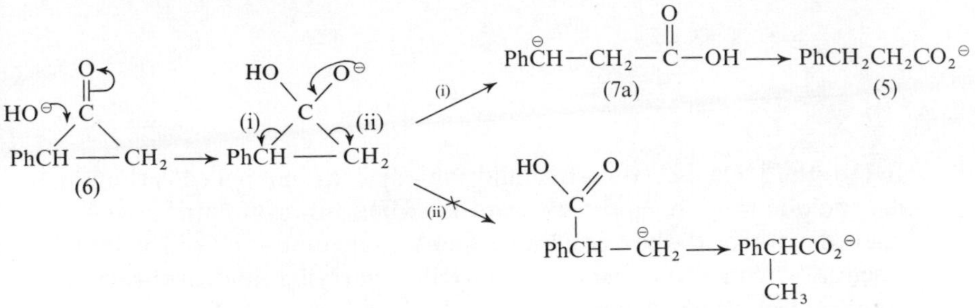

$$Ph\overset{\ominus}{C}H-CH_2-\overset{O}{\overset{\|}{C}}-OH \longrightarrow PhCH_2CH_2CO_2^{\ominus}$$
(7a)   (5)

(7b)

Direct support for the involvement of such cyclopropanone intermediates was provided by study of the rearrangement of the 2-chloro-cyclohexanone (8) in which C_1 and C_2 (i.e. the keto carbon atom and the adjacent carbon atom that carries the chlorine) are equally labelled with ^{14}C:

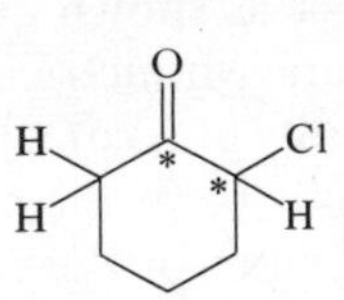

(8)

This does not mean (*cf.* p. 67) that any molecule of the chloroketone (8) contains ^{14}C labels in *both* positions, but that in the total population of molecules of (8) essentially the same number contain a *single* ^{14}C label at C_1 as contain a *single* ^{14}C label at C_2. The way in which this labelling is achieved, starting from $^{14}C_1$ labelled cyclohexanone (9)

---

*The relative stability of simple carbanions follows the sequence: tertiary < secondary < primary < benzylic.

is of some interest:

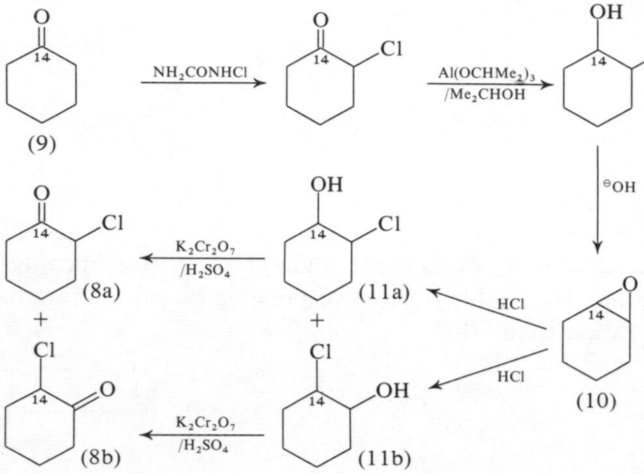

In (10) the ^{14}C labelled carbon and the adjacent unlabelled carbon in the epoxide ring are equivalent, so that when this symmetrical ring is opened with acid the resultant 1,2-chloroalcohol (11) will contain essentially equal numbers of molecules carrying hydroxyl-carbon (11a) and chlorine-carbon (11b) labels. There may, in fact, be very, very slightly more of the former as a result of a very small $k_{C^{12}}/k_{C^{14}}$ primary kinetic isotope effect (*cf.* p. 47) in going from (10) $\longrightarrow$ (11): essentially the same mixture will also be found in (8). If, then, a cyclopropanone intermediate (12) is involved in the Favorskii rearrangement of (8) with $^{\ominus}$OR, we should expect the distribution of the ^{14}C isotopic label in the product (13) to be as shown on p. 211.

As both (12a) and (12b) are symmetrical they should both undergo ring-opening with equal facility at (i) or (ii), 50 per cent of the original ^{14}C label in (8ab) should thus be in the carboxyl carbon (13a + 13a), 25 per cent in the 2-carbon atom of the cyclopentane ring (13b), and 25 per cent in the 1-carbon atom of the cyclopentane ring (13c). This is exactly the distribution that selective degradation of (13) reveals, thereby demonstrating that C_2 and C_6 (the α- and α'-positions) in the chloroketone starting material (8) must become equivalent during the course of the rearrangement, which the involvement of (12) would readily explain. To prove that the equivalence of C_2 and C_6 in (8) did not come about by, for example, migration of chlorine, the reaction was stopped short of completion and the re-isolated, as yet unrearranged, (8) shown to have exactly the same ^{14}C distribution as it had had to start with.

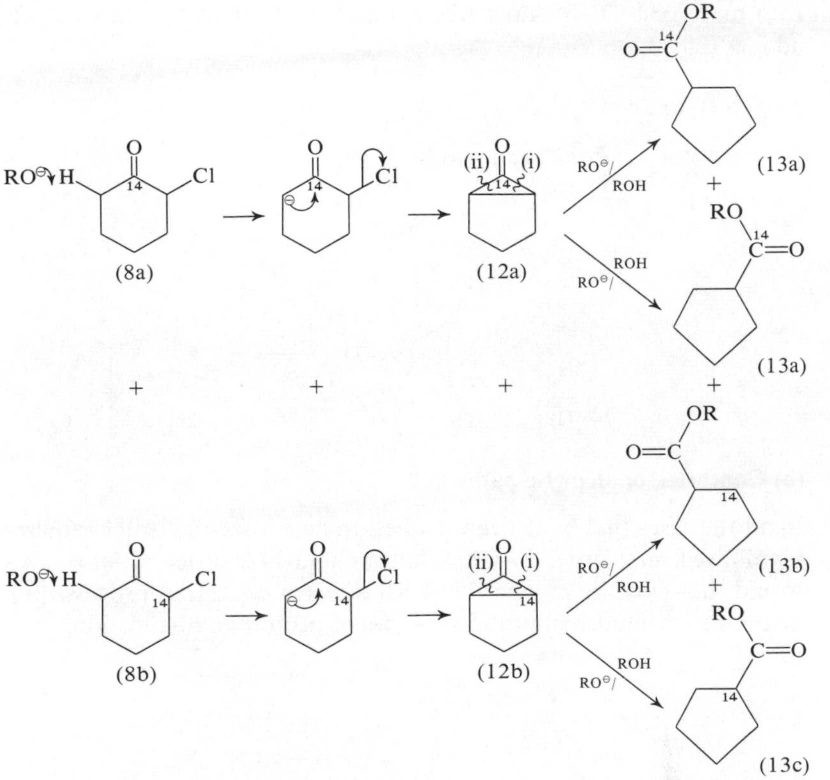

(8a) (12a) (13a) (13a) (8b) (12b) (13b) (13c)

Even more direct evidence for the involvement of cyclopropanones as intermediates is provided in the reaction of base (triethylamine) with the $\alpha\alpha'$-dibromoketone (14) by the interception of (15)—through converting it *in situ* into an isolable product, a cyclopropenone (16), that still contains the 3-membered ring:

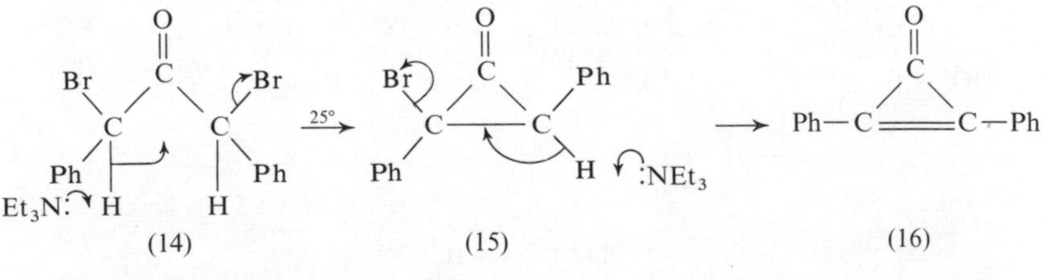

(14) (15) (16)

It has also proved possible during the attack of base (2,6-dimethylpyridine) on the chloroketone (17) to exploit the quasi double bond properties of the 3-membered ring in the intermediate (18) by inducing it to undergo a Diels–Alder like reaction with furan (19) to yield the adduct (20):

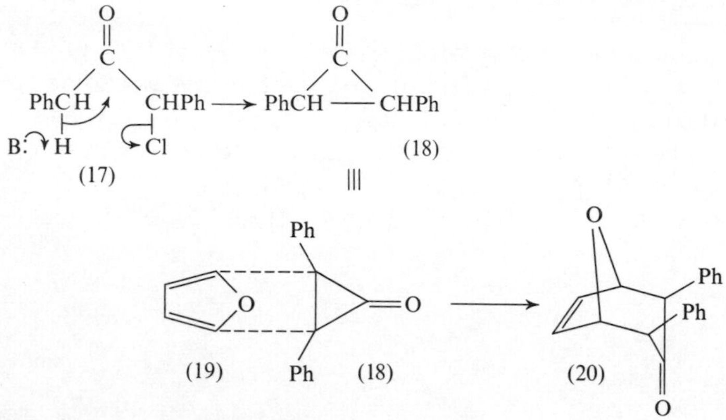

(b) Concerted or stepwise pathway?

In all the cases that have been studied to date a second order rate law is followed, first order in α-haloketone and first order in base. This would, however, be compatible with either a concerted process, [1], involving essentially simultaneous loss of proton and halide ion,

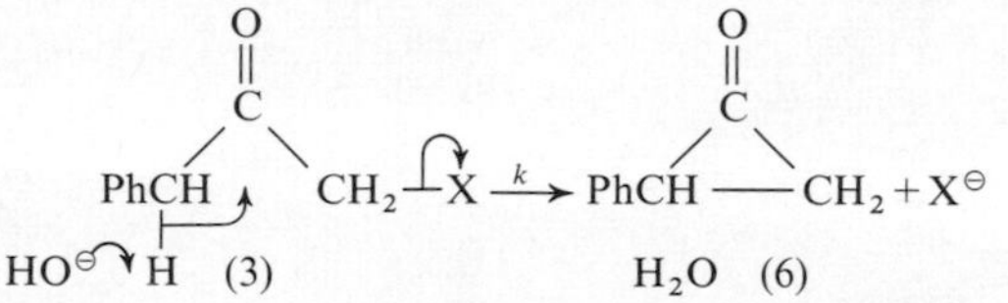

$$\text{[1]}$$

or a stepwise process, [2], involving initial formation of carbanion (21) followed by subsequent internal nucleophilic attack resulting in loss of halide ion:

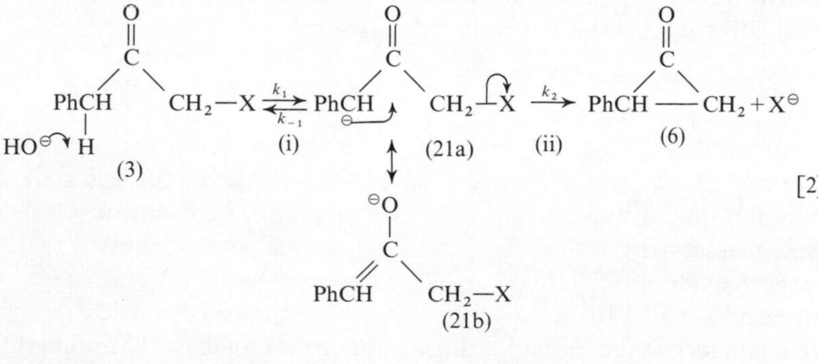

$$\text{[2]}$$

In [2] either internal attack by the carbanion (k_2) or its formation (k_1) could be rate-limiting. It is found that carrying out the reaction with base in a deuterated solvent, e.g. D_2O, MeOD, results in deuterium exchange into the benzylic (α'-) position of (3) re-isolated before the overall reaction has gone to completion. Such reversible carbanion formation could be construed as indicating the operation of step (i), i.e. k_1/k_{-1}, in the stepwise pathway, [2], thereby excluding the concerted pathway, [1]. This is not of itself a definitive test, however, as reversible carbanion (= enolate ion, 21b) formation, which clearly must occur, could merely be incidental and not integral to the reaction pathway proper, which could therefore still be a concerted one, [1].

A choice between the two alternatives can, however, be made by comparing the behaviour, under parallel conditions, of (3) and its analogue (22) that carries a methyl group on the α-carbon atom (the one attached to X):

$$
\begin{array}{c}
\text{O} \\
\|\\
\text{C} \\
\diagup \quad \diagdown \\
\text{PhCH} \quad\quad \text{CH}-\text{X} \\
| \quad\quad\quad\quad | \\
\text{H} \quad\quad\quad \text{CH}_3 \\
(22)
\end{array}
$$

If the pathway is a concerted one, [1], we would not expect the introduction of such a methyl group to have any very pronounced effect on the character of the reaction. What we find, however, is that (a) the ratio k_{Br}/k_{Cl}, the *leaving group effect*, for the pair (3, X = Br and Cl, respectively) is 63·5 whereas for the pair (22, X = Br and Cl, respectively) it is 0·9 (i.e. ≈ 1), (b) deuterium exchange into the benzylic (α'-) position of re-isolated (3, X = Cl) occurs to the extent of ≈ 80 per cent compared with ≈ 5 per cent for (22, X = Cl), and (c) the Hammett ρ value (cf. p. 171) for (3, X = Cl) is $-5·0$ compared with $+1·36$ for (22, X = Cl).

These very large differences in characteristics have no obvious explanation on the basis of a concerted pathway, [1], but they are readily explainable on a stepwise basis, [2]. Thus the extensive deuterium exchange into the benzylic (α'-) position of re-isolated (3, X = Cl) could result from [2] where $k_{-1} > k_2$, which would thus make k_2 rate-limiting. Rate-limiting attack by the carbanion carbon on the halogen-carrying (α-) carbon in (21a) would readily explain (a) the large leaving-group effect, k_{Br}/k_{Cl}, observed with (3), (b) the large negative ρ value for its reaction, and (c) the marked sensitivity of its reaction rate to the ionizing power of the solvent—change from MeOH to 50 per cent MeOH/H_2O is found to increase the reaction rate for (3, X = Cl) by a factor of 110.

By contrast, ionization of halide ion from (22, X = Cl) would be greatly accelerated by the methyl group attached to the same (α-) carbon atom, this could result in $k_2 > k_{-1}$ in [2], which would thus make k_1 rate-limiting. Rate-limiting carbanion (21a) formation with (22) would explain (*a*) the absence of a leaving-group effect, $k_{Br} \approx k_{Cl}$, (*b*) the essential lack of deuterium exchange in re-isolated (22, X = Cl), (*c*) the change in sign of the ρ value from − for (3, X = Cl) to +, and (*d*) the low sensitivity of its reaction rate to the ionizing power of the solvent—change from MeOH to 50 per cent MeOH/H$_2$O is found to increase the reaction rate for (22, X = Cl) by a factor of 6 only. There is thus detailed agreement between the experimental data and what would be expected to happen on the basis of a stepwise pathway, [2], attended by a change in the rate-limiting step on substituting (22) for (3). A closely analogous argument in favour of the stepwise pathway can also be constructed from analysis of similar experimental data derived from (23, X = Br and Cl, respectively):

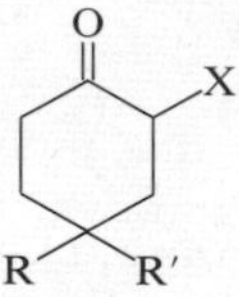

(23)

(c) Stereochemistry of reaction

We might, at first sight, expect a simple stepwise pathway, such as [2], to be attended by inversion of configuration at the α-carbon atom, the one carrying the halogen atom, i.e. (24) ⟶ (26):

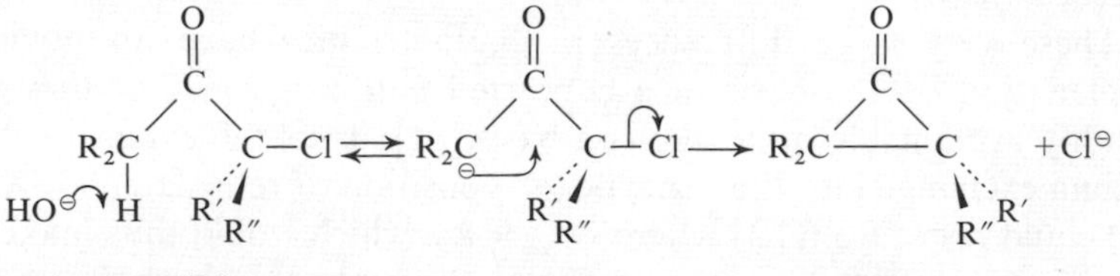

(24)   (25)   (26)

This is borne out by the observation that the *cis* chloroketone (27a) on treatment with PhCH$_2$O$^\ominus$Na$^\oplus$ in ether yielded, after acid hydrolysis of the initial benzyl ester, only the acid (28a, R = H); the isomeric *trans* chloroketone (27b) was found, similarly, to yield only the acid (28b, R = H):

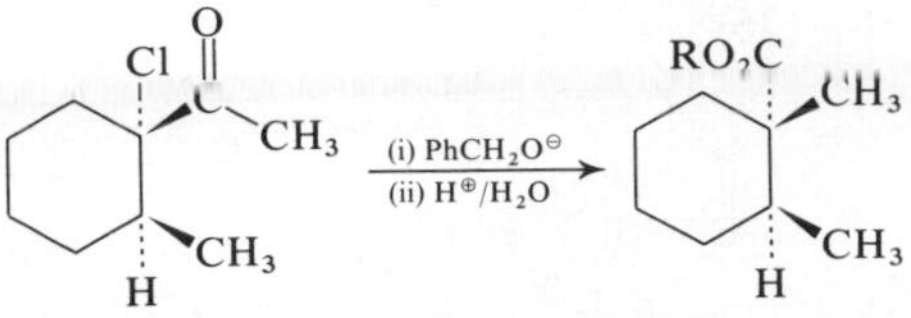

Cl
O
C
CH₃
CH₃
H
(i) PhCH₂O⊖
(ii) H⊕/H₂O
RO₂C
CH₃
CH₃
H

(27a) (28a, R = H)

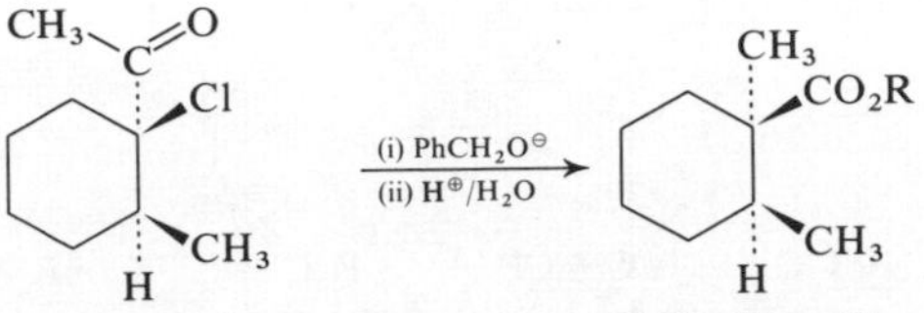

CH₃
C
O
Cl
CH₃
H
(i) PhCH₂O⊖
(ii) H⊕/H₂O
CH₃
CO₂R
CH₃
H

(27b) (28b, R = H)

The rearrangement is thus stereospecific in both cases, with inversion of configuration taking place as expected at the carbon atom from which halide ion departs. Rearrangement of (27a) with MeO⊖ in 1,2-dimethoxyethane, a solvent of low polarity, was found to result in a 95 per cent yield of the methyl ester (28a, R = Me), i.e. the reaction is again very largely stereospecific, but rearrangement of (27a) with MeO⊖ in the much more polar solvent methanol resulted in a mixture of (28a, R = Me) and (28b, R = Me), with the latter predominating:

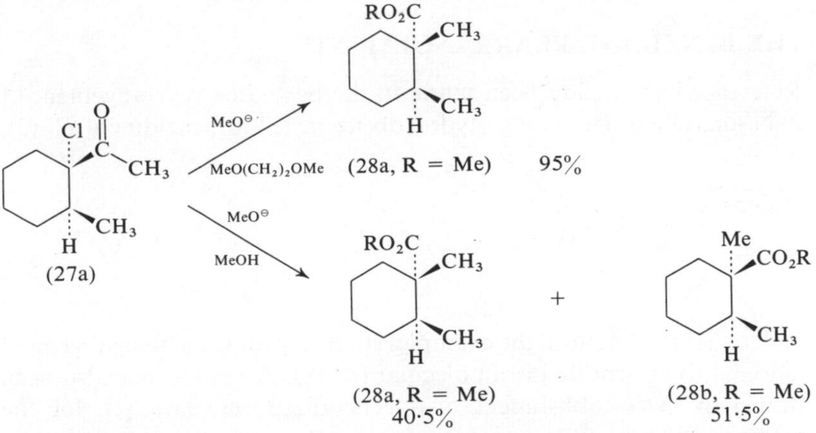

MeO⊖
MeO(CH₂)₂OMe
MeO⊖
MeOH
(27a)
(28a, R = Me)
95%
+
(28a, R = Me)
40·5%
(28b, R = Me)
51·5%

The degree of stereospecificity of the rearrangement thus depends in this case on the polarity of the solvent, and this has been observed for other examples in addition to (27a). It may be that in the more highly polar solvents a zwitterionic form (29) develops by loss of halide ion from the carbanion (25);

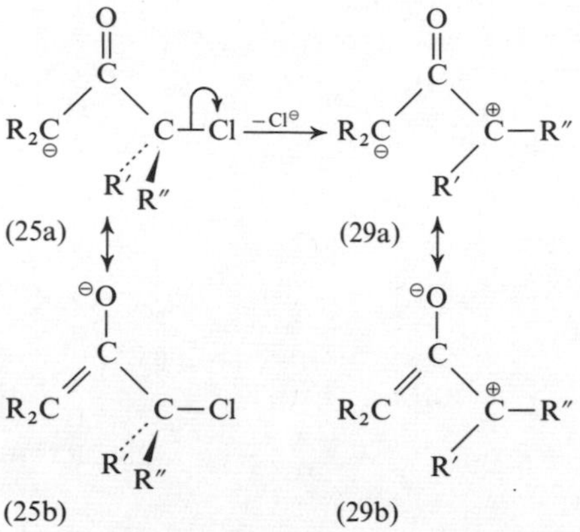

(25a)　　　　　(29a)

(25b)　　　　　(29b)

this would become planar about the new carbonium ion carbon so that any subsequent attack by carbanion carbon could take place from either side—leading ultimately to either epimer, e.g. (28a) *or* (28b). A thorough-going correlation of stereochemical with kinetic data—in particular ρ values, leaving-group effects etc.—on the same compound should prove instructive.

THE BENZIDINE REARRANGEMENT*

Reference has already been made to the benzidine rearrangement of N,N′-diarylhydrazines, e.g. hydrazobenzene (1) to benzidine itself (2),

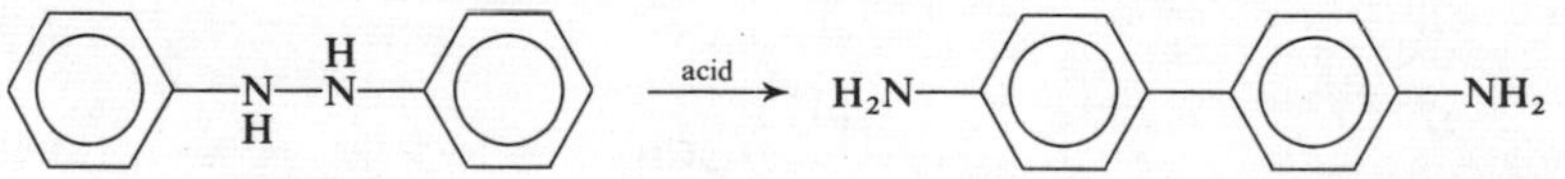

(1)　　　　　　　　　　　　　　　　(2)

and to the fact that in all the examples studied to-date the rearrangement is found to be wholly intramolecular (p. 58). Attention has also been drawn to the establishment of a generalized rate law, [1], for the rearrangement (p. 19),

$$\text{Rate} = k_1[\text{ArNHNHAr}][\text{H}^{\oplus}] + k_2[\text{ArNHNHAr}][\text{H}^{\oplus}]^2 \qquad [1]$$

*Cf. H. J. Shine, *Aromatic Rearrangements* (Elsevier, 1967), pp. 126–178 and refs. therein.

which is reflected in the non-integral rate law, [2], observed with
N,N′-di-[2-methylphenyl]hydrazine (3):

Rate $= k[\text{ArNHNHAr}][\text{H}^\oplus]^{1\cdot6}$ [2]

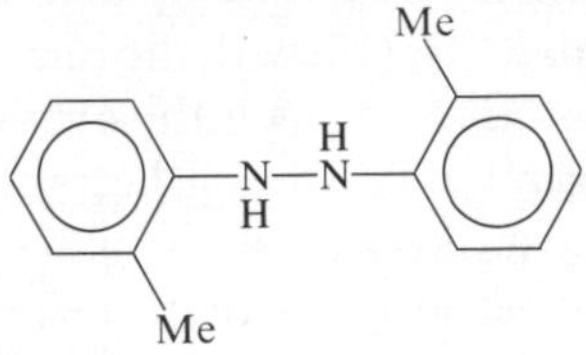

(3)

With a number of the N,N′-diarylhydrazines that exhibit such non-
integral orders in $[\text{H}^\oplus]$ at the particular acid concentration studied,
increase in acid concentration is found to result in a move towards a
rate law that is purely second order in $[\text{H}^\oplus]$, i.e. towards preponderance
of the second term in [1]; while decrease in acid concentration is found
to result in a move towards a rate law that is purely first order in $[\text{H}^\oplus]$,
i.e. towards preponderance of the first term in [1]. In a number of
cases, however, the same rate law is followed no matter what the acid
concentration; thus N,N′-di[1-naphthyl]hydrazine (4) always follows
the first order rate law in $[\text{H}^\oplus]$, [3],

Rate $= k[\text{ArNHNHAr}][\text{H}^\oplus]$ [3]

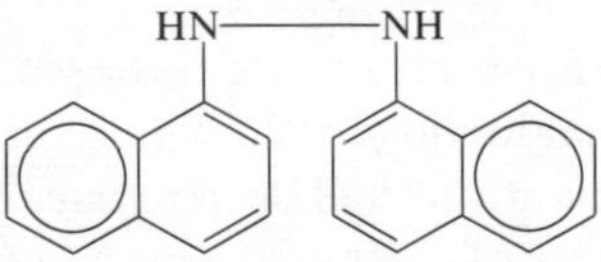

(4)

while hydrazobenzene itself (1) always follows the second order rate
law in $[\text{H}^\oplus]$, [4]:

Rate $= k[\text{ArNHNHAr}][\text{H}^\oplus]^2$ [4]

(a) The rate-limiting step

With this as a background, it is now possible to give some consideration
to the overall reaction pathway and, in particular, to seek to identify
the slow, rate-limiting step therein. Taking hydrazobenzene (1) as our
model, there are an absolute minimum of four operations that have to
be accomplished during its transformation into benzidine (2), though
some of these may be concerted with each other:

(a) Protonation—presumably of the nitrogen atoms—by two protons.
(b) Breaking of the N—N bond.
(c) Forming of the C—C bond.
(d) Loss of two protons—presumably from two carbon atoms of
the benzene rings.

To start from the end, whether or not the final proton loss, step (*d*), is rate-limiting may be determined by synthesizing the hydrazobenzene analogue (1a) with the relevant hydrogen atoms replaced by deuterium, and then seeing if the overall reaction rate, as reflected in k, is affected thereby, i.e. does it become slower? The relevant hydrogen atoms are most likely those—originally in the *p*- (4,4′-) positions of the benzene nuclei in (1a)—that are lost ultimately from adjacent carbon atoms in (6a):

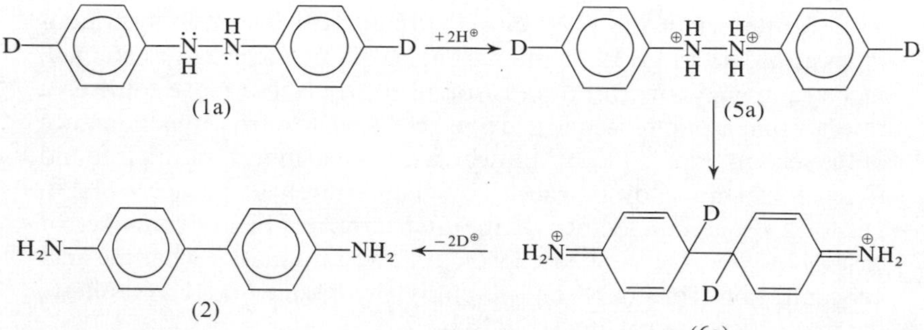

However, in case this assumption was unjustified the hydrazobenzene analogue (1b) in which *all* the hydrogen atoms *but* those in the 4,4′-positions had been replaced by deuterium was also synthesized, and its rate of rearrangement compared with that of hydrazobenzene (1) under comparable conditions. The results are as follows:

Position of D	10^3k
None	2·80
4,4′	2·85
All but 4,4′	2·87

i.e. there is no kinetic isotope effect observed, and the final loss of proton cannot therefore be the rate-limiting step.

Turning now to the initial protonation, step (*a*), there is no doubt from the observed rate law, [3], that two protons have been taken up by the time the transition state is reached, the question is whether their uptake is slow or fast:

$$\text{A). PhNHNHPh} \xrightarrow[\text{slow}]{2\text{H}^\oplus} \text{Ph}\overset{\oplus}{\text{N}}\text{H}_2\overset{\oplus}{\text{N}}\text{H}_2\text{Ph} \xrightarrow[\text{fast}]{} \text{T.S.}$$

$$\text{B). PhNHNHPh} \underset{\text{fast}}{\overset{2\text{H}^\oplus}{\rightleftharpoons}} \text{Ph}\overset{\oplus}{\text{N}}\text{H}_2\overset{\oplus}{\text{N}}\text{H}_2\text{Ph} \xrightarrow[\text{slow}]{} \text{T.S.}$$

It should be possible to distinguish between these two alternatives by carrying out the rearrangement in D_2O. If the rate-limiting (slow) step is transfer of proton from acid ($H_3O^\oplus$) to PhNHNHPh, i.e. pathway A, then use of $D_3O^\oplus$ in D_2O will lead to a *slowing* of the overall reaction rate (*cf. k*), because $D_3O^\oplus$ transfers $D^\oplus$ more slowly than $H_3O^\oplus$ transfers $H^\oplus$, under otherwise comparable conditions. If, however, the rate depends on the concentration of $Ph\overset{\oplus}{N}H_2\overset{\oplus}{N}H_2Ph$—formed in a rapidly established equilibrium before the slow step, i.e. pathway B— then the overall reaction rate will be *speeded up* in D_2O (≈ 2 fold per proton), because $D_3O^\oplus$ in D_2O is a stronger acid than $H_3O^\oplus$ in H_2O and the same concentration of the former will thus lead to a larger equilibrium concentration of $Ph\overset{\oplus}{N}H_2\overset{\oplus}{N}H_2Ph$, under otherwise comparable conditions, with consequent rise in overall reaction rate. It is found in practice (in 60 per cent dioxan/40 per cent D_2O or H_2O, with 0·19M $HClO_4$ as catalyst) that $k_{D_2O}/k_{H_2O} = 4·8$, i.e. the reaction is speeded up in D_2O, thus pathway B is operative, and the initial protonation cannot therefore be the rate-limiting step of the reaction either.

This leaves breaking of the N—N bond—step (*b*), forming of the C—C bond—step (*c*), or a concerted step involving both (*b*) and (*c*) as the possible rate-limiting step of the overall reaction; no detailed information is available in order to allow of a clear distinction between these alternatives, though opinion leans towards the concerted (*b*)/(*c*) step. It now remains to be seen if any useful information can be gleaned about the nature of the transition state. To that end it is found that the rearrangement, at a given acid concentration, is greatly accelerated by the addition of neutral salts, i.e. there is a large positive *salt effect* (*cf.* p. 31), and the rate is also found to increase as the rearrangement is carried out in solvents of progressively increasing polarity: both of these experimental observations are diagnostic of a reaction in which the transition state of the rate-limiting step is more polar than the intermediate, or other species, that precedes it, i.e. (5):

$$(5) \qquad \xrightarrow[\text{slow}]{} \text{T.S.} \longrightarrow (6)$$

(b) The polar transition state

The intermediate (5) is itself highly polar so that the transition state must therefore be extremely highly polar, and the suggestion has been made that initiation of heteropolar fission of the N—N bond in (5) leads to a state (7) in which both positive charges are becoming concentrated on one aromatic moiety, leaving the other with less and less of its original positive charge:

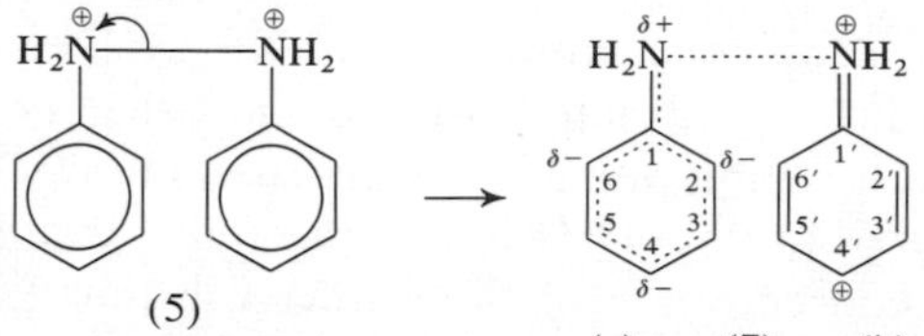

(5)

(a)   (7)   (b)

The initial driving force for the N—N bond fission springs from the proximity of the two positive charges on the adjacent nitrogen atoms, but it may well be objected that in a symmetrical species such as (5) homopolar fission of the N—N bond to yield radical cations (8) would be the preferred mode:

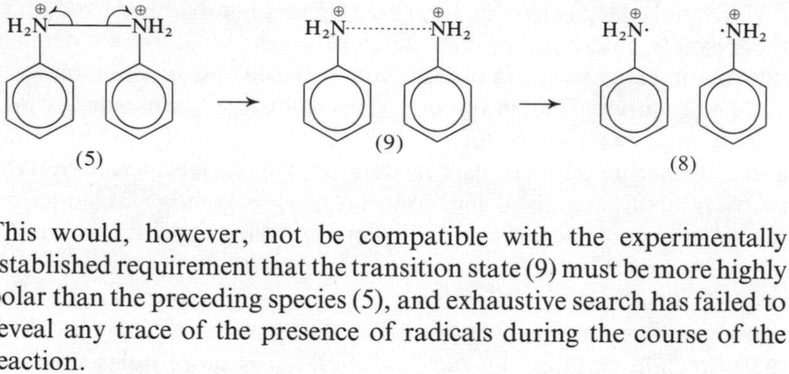

(5)

(9)

(8)

This would, however, not be compatible with the experimentally established requirement that the transition state (9) must be more highly polar than the preceding species (5), and exhaustive search has failed to reveal any trace of the presence of radicals during the course of the reaction.

The species (7) above has been represented designedly with only bond-breaking shown so that, from the most likely distribution of developing charge, it may be possible to forecast the site or sites at which the new bond will form. The moiety (7a) is essentially of an aniline type and would be expected to exhibit the charge distribution as shown; the bipolar cation moiety (7b) has to accommodate two positive charges and these are likely to be as widely separated as possible, thus leading to concentration of positive charge in the *p*- (4'-) position. This position will thus be enormously electrophilic, and we should therefore expect it to attack the activated 4- and 2-positions of the aniline-like moiety (7a), the former predominating because of its proximity:

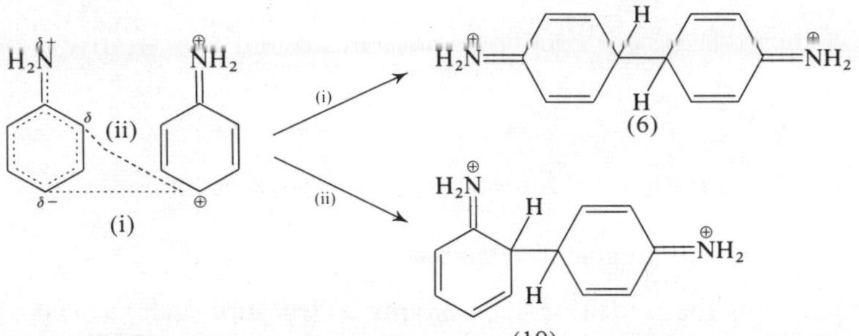

This is precisely what is observed in practice; a typical rearrangement of hydrazobenzene (1) is found to result in the formation of ≈ 70 per cent benzidine (2, derived from (6), *cf.* above), and ≈ 30 per cent diphenyline (11) derived from (10):

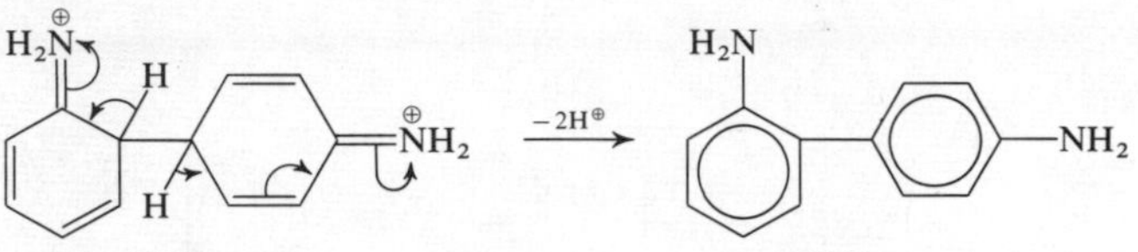

In the above discussion, N—N bond-breaking and C—C bond-forming have, for simplification of the argument and clarity of representation, been represented as separate operations, but the chances are that in practice they occur essentially simultaneously in a concerted operation.

(i) Structure and reaction rate:

If the transition state does indeed resemble (7), we should expect any factor that served to stabilize the bipolar moiety (7b) to be of determining significance in its formation. Such a factor would be the presence of an electron-donating substituent for this, by delocalizing the charge on the bipolar moiety (12b),

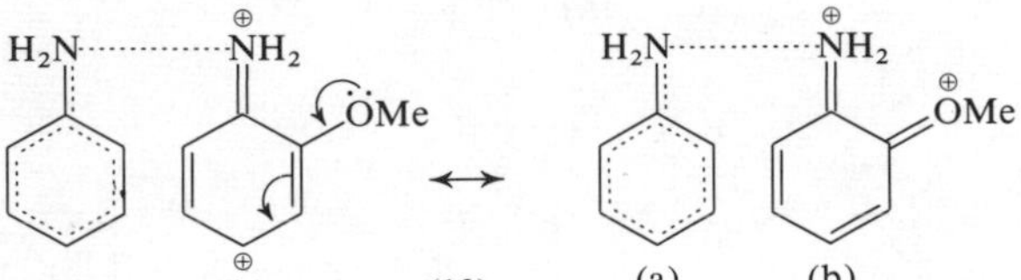

could lower the latter's energy level, and hence increases the rate at which it would be formed, i.e. the rate of the overall reaction. This, too, is what is observed experimentally as the following k values demonstrate:

Ar	Ar'	$10^3 k$
C_6H_5	C_6H_5	6.87×10^{-3}
C_6H_5	2-MeOC_6H_4	1.54
2-MeOC_6H_4	2-MeOC_6H_4	Too fast to measure

(ii) Structure and product distribution:

Evidence for the transition state having a structure such as (7) is also provided by a detailed study of the relative distribution of the products that are formed on rearrangement of substituted hydrazobenzenes. If a hydrazobenzene carries two *p*- (4,4'-) substituents (13), then neither a benzidine (e.g. 2) nor a diphenyline (e.g. 11) can be formed, and the product of rearrangement is found to be an *o*-semidine (14):

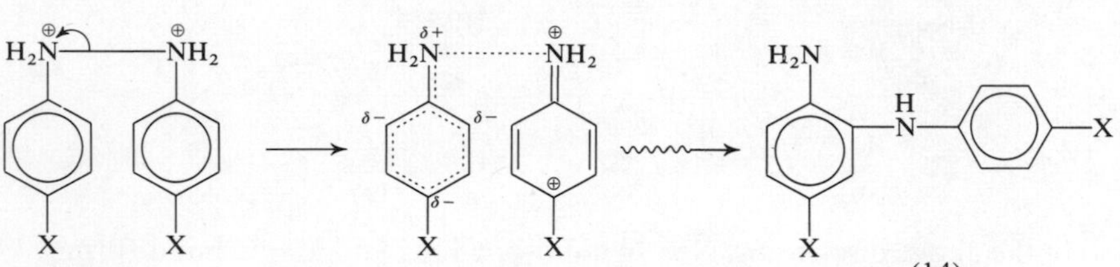

(13)   (14)

Where the 4- and 4'-substituents in the hydrazobenzene are not the same (15), there is the possibility of obtaining two alternative *o*-semidines (16 and 17) depending on which moiety in the transition state (18b *v*. 19b) assumes the bipolar state:

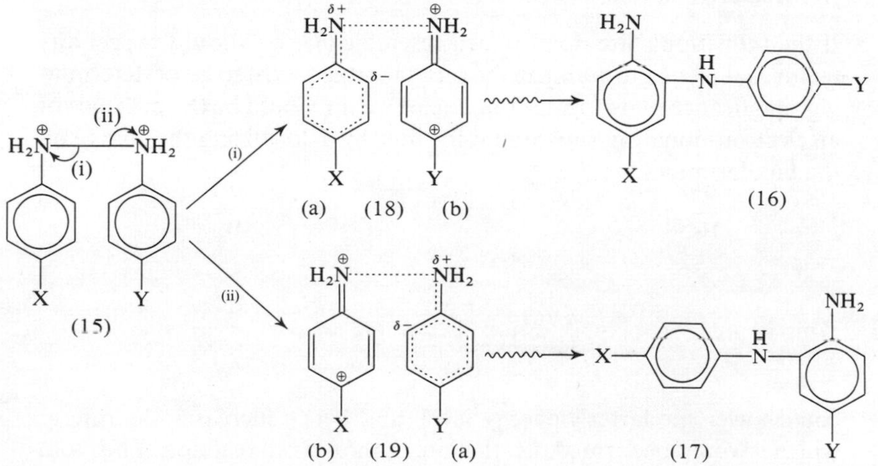

(a)   (18)   (b)   (16)

(15)

(b)   (19)   (a)   (17)

Where X and Y are closely similar in nature we should expect to get both products, as indeed commonly happens. If, however, one substituent is very much more electron-donating than the other we should expect that transition state to preponderate in which this substituent is in the bipolar (*b*) moiety, because of the greater stabilization of the latter that is thereby effected. This will, of course, lead to the *o*-semidine in which the highly electron-donating substituent is *p*- to the NH group, and this should therefore be the major, or conceivably, the sole product. Thus with the hydrazobenzene (20),

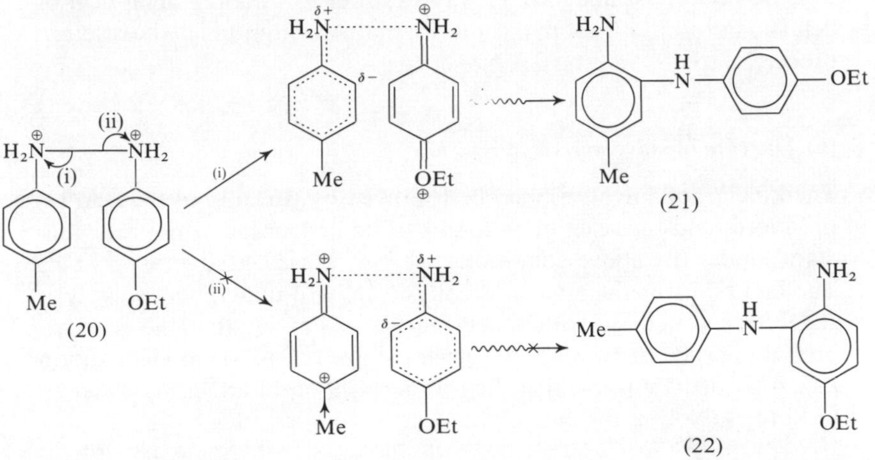

(21) is found to be all but the exclusive product of rearrangement.

The benzidine rearrangement in all its ramifications is a good deal more complex than the above consideration would suggest; several alternative models have been suggested for the transition state, but on the evidence available to date a highly polar species such as (7) best fits the experimental evidence. A good deal still remains to be elucidated, however, despite over a hundred years of study, and the polar transition state model (7) may well require subsequent modification.

INTERNAL ION PAIR RETURN IN THE SOLVOLYSIS OF AN ALLYLIC ESTER*

This is an extremely ingenious investigation of the solvolysis of an allylic ester—in fact of *cis* 5-methyl-2-cyclohexenyl *p*-nitrobenzoate (1)

*H. L. Goering, J. T. Doi, and K. D. McMichael, *J. Amer. Chem. Soc.*, 1964, **86**, 1951 and refs. therein.

in 80 per cent aqueous acetone at 100°—

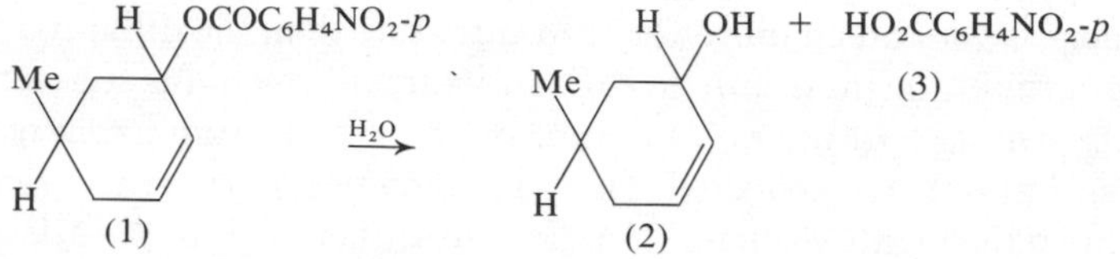

in which kinetic, stereochemical, and ^{18}O labelling techniques are used together in a most imaginative way in order to obtain a great deal of detailed information about the intermediate through which the reaction proceeds.

(a) The rate of solvolysis, $k_{sol.}$

The kinetics of solvolysis may be followed by titration of the acid (3) produced, and the reaction is found to be first order with a rate constant, under the above conditions, of $k_{sol.} = 1.37 \pm 0.04 \times 10^{-2}$ hr^{-1}. The fact that it is the ester of an allylic alcohol undergoing solvolysis suggests, by analogy with other known examples, that the reaction probably proceeds by alkyl/oxygen cleavage (*cf.* p. 51) to yield an ion pair intermediate (4), which then undergoes rapid attack by water to yield the solvolysis product (2):

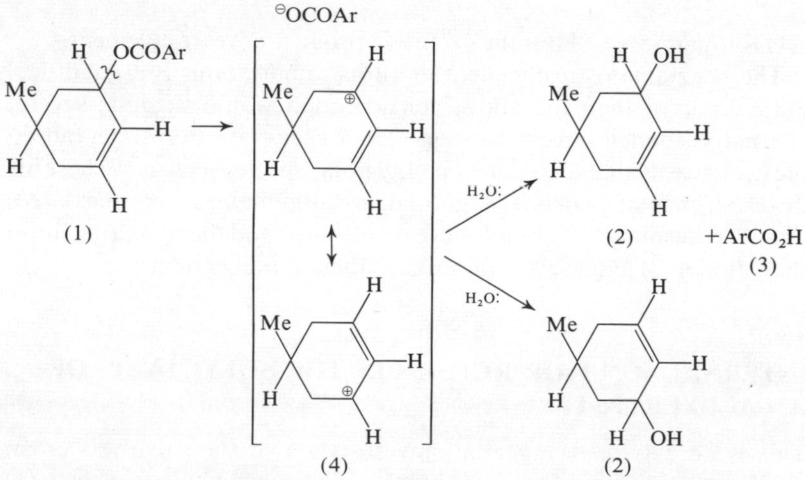

Such a reaction pathway is, of course, entirely compatible with the above kinetic measurements; the stabilization by delocalization that can occur in both the allylic cation and the carboxylate anion moieties of (4) would be expected to promote such a reaction pathway, at least in so far as the transition state preceding the ion pair intermediate (4) resembles it in nature. Direct supporting evidence for alkyl/oxygen cleavage in (1) is provided by the fact that the allylic ester is found to solvolyse 240 times faster, under comparable conditions, than the corresponding saturated—non-allylic—cyclohexyl ester, which is known to undergo solvolysis by acyl/oxygen cleavage (*cf.* p. 51). There is no valid reason why introduction of an allylic double bond should speed acyl/oxygen cleavage to anything like this extent, whereas a change of mechanism to alkyl/oxygen cleavage would account for it admirably. It is further found that the rate of solvolysis of (1) is increased very markedly as the solvent is made more polar: as would be required by the highly polar transition state that must precede (4).

(b) The rate of loss of optical activity, $k_{rot.}$

The *cis* ester (1) is a racemate ($\pm$) and can therefore be resolved into ($+$) and ($-$)enantiomers. If the solvolysis is repeated on one of these resolved, optically active, forms, e.g. ($-$)(1a), it is found that all optical activity is lost, the product allylic alcohol (2ab) being a racemate ($\pm$). This fact is readily explained by the involvement of an ion pair intermediate, i.e. (4), for its allylic carbonium ion moiety will be symmetrical:

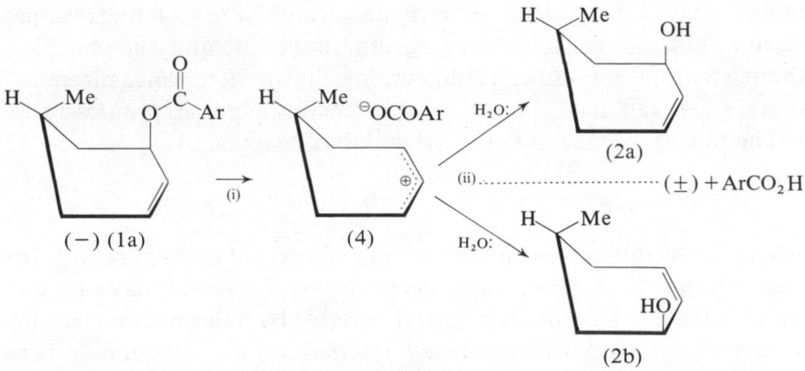

*These, as written, are somewhat distorted structures but illustrate the particular stereochemical point at issue rather more readily than do the more correct structures such as:

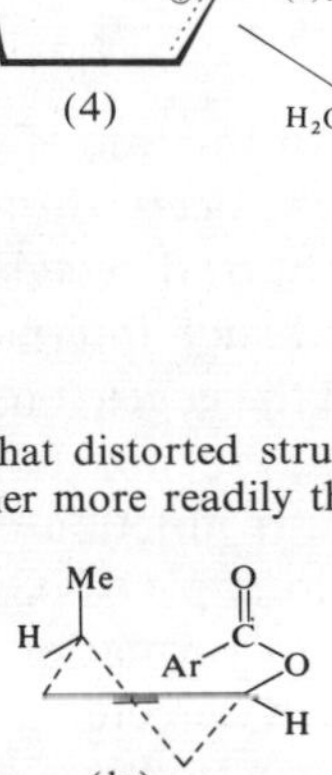

(1a)

Attack by H_2O: on one allylic carbon atom will lead to alcohol (2a), while attack on the other will lead to alcohol (2b); it may readily be seen that these are mirror images of each other. As the carbonium ion is symmetrical, attack on the carbon atom at one end of its allylic system is just as likely as attack on the carbon atom at the other, equal quantities of (2a) and (2b) will thus be obtained, i.e. a racemate (2ab). The first order rate constant, $k_{rot.}$, for loss of optical activity by (1a), measured polarimetrically, is found to have a value of $2\cdot39 \pm 0\cdot04 \times 10^{-2} hr^{-1}$.

(c) The rate of racemization of original ester, $k_{rac.}$

It will thus be seen that the value of $k_{rot.}$ is larger than that of the first order rate constant, $k_{sol.}$, for the solvolysis of (1a) under comparable conditions: the rate of loss of optical activity by (1a) is thus faster than the rate at which it undergoes solvolysis, and the former must therefore occur by some process *in addition to* its conversion into the racemic alcohol (2ab). One possible additional process could be the reversal of step (i) above: attack on the allylic carbonium ion in (4) by its gegen ion, $ArCO_2^{\ominus}$—rather than by H_2O: in step (ii)—resulting in its reconversion to starting material (1). Just as with H_2O:, attack could take place on either allylic carbon in the cation to yield (1a) and (1b), respectively; the result would thus be racemization of the starting material (1a) by what is called *internal ion pair return*. That this is indeed what is taking place can be demonstrated by re-isolating residual starting material (1) after varying time intervals, and showing that though the total amount of it is decreasing, the extent to which racemization (1a⟶ 1ab) has occurred in what remains increases progressively.

The rate of racemization of (1a) will thus be given by,

$$k_{rac.} = k_{rot.} - k_{sol.} = 1\cdot02 \pm 0\cdot05 \times 10^{-2} \ hr^{-1}$$

where $k_{rac.}$ is thus a measure of the rate of internal ion pair return. Ion pair return will, of course, also occur during the solvolysis of $(\pm)(1)$, i.e. (1ab), but will naturally not be detectable by polarimetric measurements. That the occurrence of such internal ion pair return reflects an extremely close association of the gegen ions in the ion pair intermediate (4) is witnessed by the fact that racemization of (1a)—ion pair return— is not accompanied by the introduction of any ^{18}O label when ^{18}O labelled *p*-nitrobenzoate anion, $p\text{-}O_2NC_6H_4C^{18}O_2^{\ominus}$, is introduced into the solution. Further, no formation of the geometrical (*trans*) isomeride of (1a) has ever been detected as a result of ion pair return, indicating that the carboxylate anion is confined, in the ion pair (4), in close proximity to that face of the forming, locally planar carbonium ion from which it originally departed.

(d) The rate of ^{18}O equilibration in ion pair return, k_{equil}.

The ion pair return was now studied further by solvolysing, under comparable conditions, the allylic ester (1c) in which the carbonyl oxygen atom was ^{18}O labelled (*cf.* p. 51). After varying time intervals the residual—as yet unsolvolysed—ester was reisolated and examined to see whether the ^{18}O label remained confined to the carbonyl oxygen atom (1c), or whether some was now present in the ether oxygen atom of the ester (1d), i.e. whether any *equilibration* of the label had taken place:

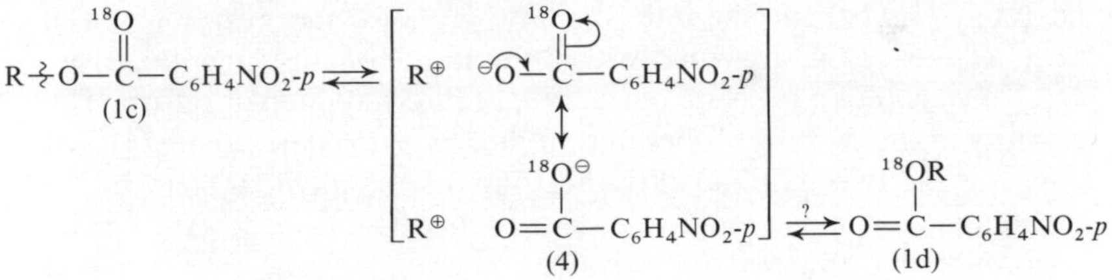

To establish how much, if any, ^{18}O label was present in the ether oxygen atom of the ester, i.e. (1d), the re-isolated, unsolvolysed ester (1c + ?1d) was hydrolysed with base, which is known to involve purely acyl/oxygen cleavage. The resultant substituted cyclohexenol (2c + ?2d) was then separated from the carboxylate anion and converted, for convenience, into the solid *p*-nitrobenzoate ester (1e + ?1f); an ^{18}O determination (*cf.* p. 52) was then carried out on this:

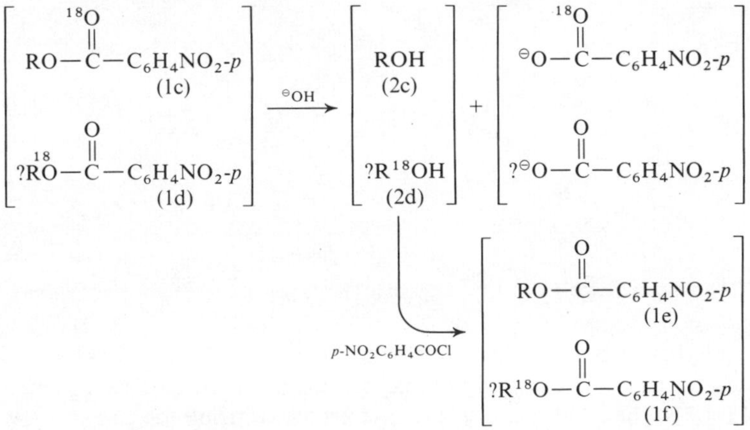

Any ^{18}O label detected in (1e + ?1f) will thus be the ^{18}O label that was present in the *ether* oxygen of the original, re-isolated, unsolvolysed ester, i.e. (1d) in (1c + ?1d). The above sequence of operations was repeated on unsolvolysed ester (1c + ?1d) re-isolated after successive time intervals—a limit is, of course, set by the ultimate total solvolysis of (1c + ?1d)—and equilibration of the label (carbonyl oxygen ⟶ ether oxygen) was indeed found to have taken place. The equilibration was found to follow a first order rate law with a rate constant, $k_{equil.}$, having a value of $0.97 \pm 0.03 \times 10^{-2}$ hr^{-1}.

(e) Two limiting ion pair intermediates

It will thus be seen that the rate constant, $k_{equil.}$, for equilibration in (1), i.e. (1c) ⟶ (1d) and the rate constant, $k_{rac.}$, for racemization of (1), i.e. (1a) ⟶ (1ab), are equal within the limits of experimental error: $k_{equil.}$ also is thus a measure of the rate of internal ion pair return. The equality of the two establishes that in the ion pair intermediate (4), not only are the two terminal carbon atoms of the allylic system in the carbonium ion moiety equivalent to each other (a necessary consequence of the observed racemization above), but the two carboxyl oxygen atoms of the carboxylate ion moiety are also equivalent to each other (a necessary consequence of the observed ^{18}O label equilibration above).

It now remains to seek information on the relationship between the pair of mutually equivalent allylic carbon atoms and the pair of mutually equivalent carboxylate oxygen atoms in the ion pair intermediate (4); possible alternative relationships can be expressed diagrammatically as follows:

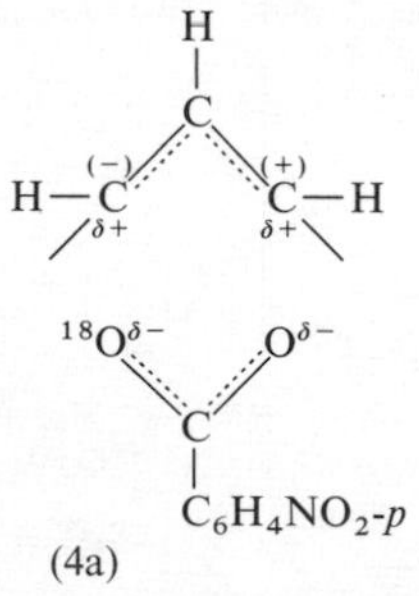

(4a) 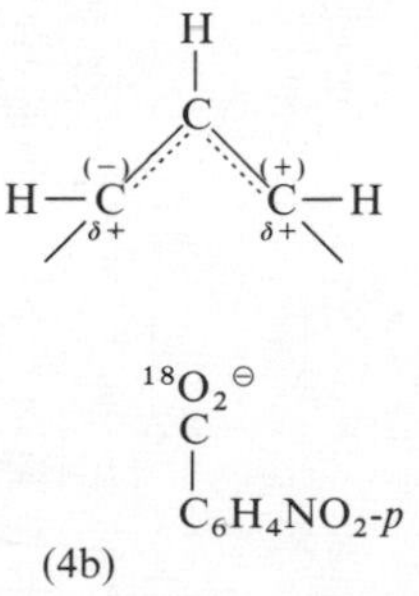   (4b)

The allylic carbon atoms are marked (+) and (−) to indicate that attack on the (+)carbon by an oxygen atom of the carboxylate anion will lead to the (+)form of (1), while attack on the other, i.e. (−), will lead to the $(-)$form of the ester. It should be emphasized that (4a) and (4b) are not intended to express any specific spatial connotation, but merely to indicate, diagrammatically, the two possible limiting cases so far as the relation between the pair of carbon and the pair of oxygen atoms are concerned. The only spatial comment that can be made is that the carboxylate anion remains wholly associated in (4) with the face of the locally planar allylic carbonium ion from which it originally departed, as we saw above (p. 226). Formulation (4a) is intended to convey that the allylic carbon atoms and the carboxylate oxygen atoms are *specifically paired* with each other: we should thus expect attack by the ^{18}O labelled carboxylate oxygen atom to happen, in the extreme case represented by (4a), exclusively on the $(-)$ allylic carbon atom, and attack by the non-labelled oxygen atom exclusively on the $(+)$ allylic carbon atom. By contrast, formula (4b) is intended to convey that the carboxylate oxygen atoms are equivalent with respect to *each* allylic carbon atom: we should thus expect attack by the ^{18}O labelled or the non-labelled oxygen atom to happen with equal ease on either allylic carbon atom.

(f) Choosing between the limiting cases

For examples, such as the present one, where $k_{rac.} = k_{equil.}$ it is possible to distinguish between the alternative limiting cases, (4a) and (4b), by investigating internal ion pair return, under the same conditions as have been employed to date, in an optically pure, discretely ^{18}O labelled ester, and looking for the occurrence or non-occurrence of randomization of the ^{18}O label in *each* of the resultant enantiomers. Thus starting with $(-)R^{18}OCOC_6H_4NO_2$-p (1g), the limiting case (4a) for the ion pair intermediate would require internal ion pair return to occur with equal facility *either* by attack of ^{18}O on $(-)C$ leading to reformation of (1g) *or* by attack of O on $(+)C$ leading to (1h):

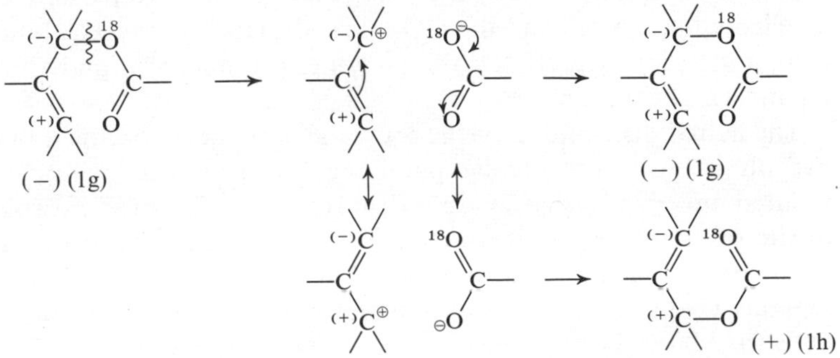

Thus *no* randomization of the label should take place, all the ^{18}O labelled ether oxygen should be in the $(-)$enantiomer (1g) and all the ^{18}O labelled carbonyl oxygen is the $(+)$enantiomer. Starting with (1g) once again, the limiting case (4b) for the ion pair intermediate would require internal ion pair return to occur equally easily by attack of *either* oxygen atom (^{18}O or O) on *either* carbon atom $((+)$ or $(-))$:

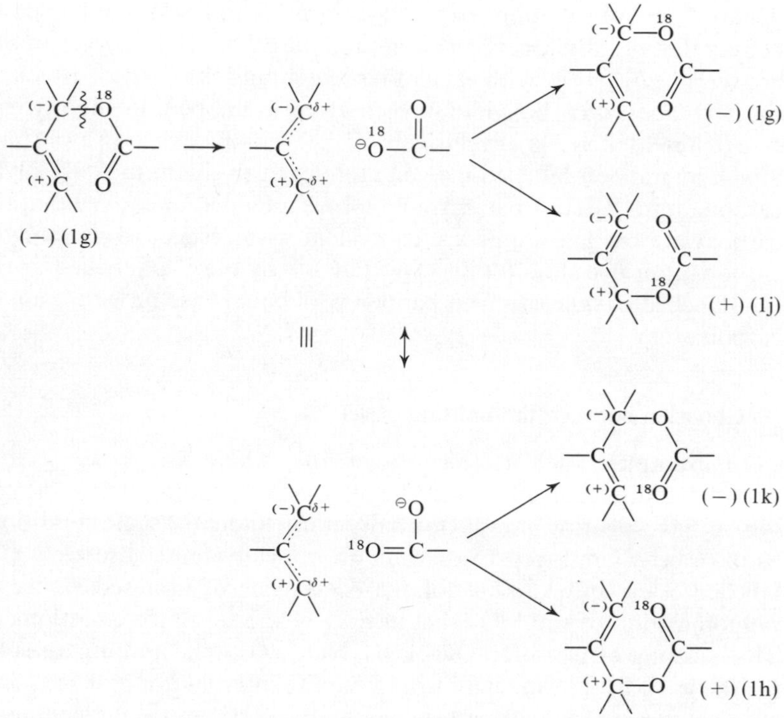

$(-)$ (1g)

$(-)$ (1g)

$(+)$ (1j)

$(-)$ (1k)

$(+)$ (1h)

Thus *complete* randomization of the label should take place, the ^{18}O labelled ether oxygen and the ^{18}O labelled carbonyl oxygen should be equally distributed between the $(-)$, i.e. (1g and 1k), and $(+)$, i.e. (1j and 1h), enantiomers.

The actual distribution of the ^{18}O label may be determined by base hydrolysis (known to involve purely acyl/oxygen cleavage) of the re-isolated, unsolvolysed ester (1g + 1h + ?1j + ?1k), followed by isolation of the resultant $(\pm)$ substituted cyclohexenol (2), conversion of this to the crystalline $(\pm)$ acid phthalate (6), resolution of the latter into separate $(-)$ and the $(+)$acid phthalates, and finally ^{18}O determinations (p. 52) on the separate $(-)$ and $(+)$ compounds:

$$
\left[
\begin{array}{l}
\overset{\displaystyle O}{(\pm)R^{18}O\!\overset{?}{\!-\!}\overset{\|}{C}\!-\!C_6H_4NO_2\text{-}p} \\
\qquad (?1j+1g) \\[2mm]
\overset{\displaystyle {}^{18}O}{(\pm)RO\!\overset{?}{\!-\!}\overset{\|}{C}\!-\!C_6H_4NO_2\text{-}p} \\
\qquad (1h+?1k)
\end{array}
\right]
\xrightarrow{{}^{\ominus}\underline{OH}}
\left[
\begin{array}{l}
(\pm)R^{18}OH \\[6mm]
(\pm)ROH \\
\qquad (2)
\end{array}
\right]
+
\left[
\begin{array}{l}
\overset{\displaystyle O}{{}^{\ominus}O\!-\!\overset{\|}{C}\!-\!C_6H_4NO_2\text{-}p} \\[2mm]
\overset{\displaystyle {}^{18}O}{{}^{\ominus}O\!-\!\overset{\|}{C}\!-\!C_6H_4NO_2\text{-}p}
\end{array}
\right]
$$

$$
\left[
\begin{array}{l}
\overset{\displaystyle O}{(\pm)R^{18}O\!-\!\overset{\|}{C}\!-\!C_6H_4CO_2H} \\[2mm]
\overset{\displaystyle O}{(\pm)RO\!-\!\overset{\|}{C}\!-\!C_6H_4CO_2H} \\
\qquad (6)
\end{array}
\right]
\xrightarrow{\text{Resolution}}
\left[
\begin{array}{l}
\overset{\displaystyle O}{(+)R^{18}O\!-\!\overset{\|}{C}\!-\!C_6H_4CO_2H \ ?(6j)} \\[2mm]
\overset{\displaystyle O}{(+)RO\!-\!\overset{\|}{C}\!-\!C_6H_4CO_2H \ (6h)}
\end{array}
\right]
$$

$$
\left[
\begin{array}{l}
\overset{\displaystyle O}{(-)R^{18}O\!-\!\overset{\|}{C}\!-\!C_6H_4CO_2H \ (6g)} \\[2mm]
\overset{\displaystyle O}{(-)RO\!-\!\overset{\|}{C}\!-\!C_6H_4CO_2H \ ?(6k)}
\end{array}
\right]
$$

The ^{18}O label values thereby obtained will, of course, be the ^{18}O alkyl oxygen for the $(-)$ and $(+)$components in the re-isolated, unsolvolysed ester, i.e. (1g) and ?(1j), respectively; knowing the total carboxyl ^{18}O content of the re-isolated ester we can now, by difference, calculate the ^{18}O carbonyl oxygen values for the $(-)$ and $(+)$components in the re-isolated, unsolvolysed ester, i.e. ?(1k) and (1h), respectively.

It is found in practice that there is, within experimental error, complete randomization of the ^{18}O label in each of the enantiomers produced by internal ion pair return, i.e. there is found to be equal sharing of the label between alkyl and carbonyl oxygen atoms in both $(-)$ and $(+)$enantiomers (note, however, that a correction has to be made for that part of the original ester (1g) that has not yet undergone racemization). The randomization is found to follow a first order rate law with a rate constant, $k_{\text{rand.}}$, having a value of $0.96 \pm 0.06 \times 10^{-2}\text{hr}^{-1}$! The rate of randomization of ^{18}O label in (1) is thus, like rate of ^{18}O equilibration ($k_{\text{equil.}}$, *cf.* p. 227) and rate of racemization ($k_{\text{rac.}}$, *cf.* p. 226), also a measure of the rate of internal ion pair return.

The ion pair intermediate, (4), involved in the solvolysis of, and internal ion pair return in, the *cis* substituted-cyclohexenyl *p*-nitrobenzoate (1) thus corresponds exactly to the limiting case (4b). Other closely analogous allyl esters are known, however, in which ^{18}O label randomization in the separate $(-)$ and $(+)$ enantiomers of unsolvolysed material is incomplete—$k_{rand.} < k_{rac.}$—and these must involve ion pair intermediates somewhere in between the limiting cases (4b) and (4a); $k_{rand.}$ for (4a) is, of course, equal to zero. Apart from the close association of the carboxylate gegenion with that face of the locally planar carbonium ion from which it originally departed (p. 226), little is known about the detailed spatial arrangement in these ion pair intermediates: a continuing challenge is thus presented even yet.

Index

236   *Index*